Acclaim for *Dragon Hunter*

"Amazing stuff . . . Andrews was a great man who led an incredibly exciting life, and you keep on reading, wondering why they don't make them like that anymore." —*National Geographic Adventure*

"Retold with verve and dash . . . the stories suggest that if Andrews and his adventures were not the direct inspiration for the Indiana Jones films, it is only because he had become so much of an archetypal figure that Indiana Jones would have to resemble him."
—*The Boston Globe*

"Reads like an action movie. . . . Post–World War I America was captivated by the daring of men like Charles Lindbergh and Richard Peary. Andrews was no less bold."
—*Forbes*

"Captivating . . . In the 1920s, Roy Chapman Andrews's name was as well known as Muhammad Ali's is today. He was the stuff of legends — a swashbuckling explorer who journeyed into the last blank left on the globe, cheated death with aplomb and unearthed some of the greatest fossil treasure troves of all time."
—*Ann Arbor News*

"Fascinating . . . the account of Andrews's exploits leaves the armchair explorer throbbing." —*Business Week*

"Lively and fast-paced . . . Gallenkamp has told the tale in a very readable and entertaining fashion." —*The New Mexican*

"A precious source of information . . . Andrews's Central Asiatic expeditions of the 1920s have taken on an almost mythical dimension."
—*Nature*

"A page-turning adventure story." —*Publishers Weekly*

"Compelling . . . Andrews was handsome, dapper, social and self-promoting. He was Indiana Jones, as played by Cary Grant."
— *Houston Chronicle*

"Andrews emerges from these pages as a hero of the modern era."
— *The New Orleans Times-Picayune*

"An exciting biography . . . *Dragon Hunter* shows how Andrews mixed scientific investigation with swashbuckling adventure."
— *The Wall Street Journal*

"Andrews contended with bandits, corrupt officials, invading armies, disease and other dangers. After finishing Gallenkamp's vigorous book, readers will understand why Andrews should have served as the model for the movie character Indiana Jones—who, if anything, pales by comparison to the real thing." — Gregory McNamee, Amazon.com

"What an exciting life Roy Chapman Andrews lived. . . . His accomplishments were notable; his adventures, unbelievable. . . . Written with verve and attention to detail, Gallenkamp's fast-paced book does justice to Andrews' eventful life."
— *Tampa Tribune and Times*

"Enormously entertaining . . . On his first journey to East Asia in 1909, when he was twenty-five years old, Andrews spent two weeks stranded on a desert island; fended off sharks after his boat was capsized by a finback whale; survived typhoons, heatstroke, poisoned bamboo stakes, headhunters and 20-foot pythons. . . . He also sampled opium; befriended Mother Jesus, Yokohama's most famous madam; enjoyed the pleasures of Shimonoseki, 'the hardest-drinking port in the East' and, along the way, collected 50 mammals, 424 birds and a new species of ant. And that's in just the first 35 pages of this propulsive, nonstop biography." — *The New York Times Book Review*

PENGUIN BOOKS

DRAGON HUNTER

A writer and archaeologist, Charles Gallenkamp is the author of *Maya: The Riddle and Rediscovery of a Lost Civilization*. He served as coordinator of the highly acclaimed exhibition "Maya: Treasures of an Ancient Civilization," which toured the United States and Canada from 1985–1987. He lives in Sante Fe, New Mexico, with his wife Karen L. Wright.

Michael J. Novacek is senior vice president, provost of science, and curator of paleontology at the American Museum of Natural History. He is the author of *Dinosaurs of the Flaming Cliffs*.

DRAGON

HUNTER

Roy Chapman Andrews and the
Central Asiatic Expeditions

CHARLES GALLENKAMP

Foreword by Michael J. Novacek

In Association with the American Museum of Natural History

PENGUIN BOOKS

For Karen, with gratitude and love

PENGUIN BOOKS
Published by the Penguin Group
Penguin Putnam Inc., 375 Hudson Street, New York, New York 10014, U.S.A.
Penguin Books Ltd, 80 Strand, London WC2R 0RL, England
Penguin Books Australia Ltd, 250 Camberwell Road,
Camberwell, Victoria 3124, Australia
Penguin Books Canada Ltd, 10 Alcorn Avenue, Toronto, Ontario, Canada M4V 3B2
Penguin Books India (P) Ltd, 11 Community Centre,
Panchsheel Park, New Delhi-110 017, India
Penguin Books (N.Z.) Ltd, Cnr Rosedale and Airborne Roads,
Albany, Auckland, New Zealand
Penguin Books (South Africa) (Pty) Ltd, 24 Sturdee Avenue,
Rosebank, Johannesburg 2196, South Africa

Penguin Books Ltd, Registered Offices: Harmondsworth, Middlesex, England

First published in the United States of America by Viking Penguin,
a member of Penguin Putnam Inc. 2001
Published in Penguin Books 2002

1 3 5 7 9 10 8 6 4 2

Drawings by Karen L. Wright

Photographs of Andrews at fourteen and twenty are courtesy of the Beloit Historical
Society, Beloit, Wisconsin; photograph of Yvette Andrews and her son George is courtesy
of George B. Andrews; photograph of Andrews on the cover of *Time* is courtesy of TimePix;
photographs of Andrews playing polo and of his collection of Chinese art in their
New York City apartment are courtesy of Mrs. Robert A. Street.

All other photographs and the map on pp. xx–xxi are courtesy of
the American Museum of Natural History.

THE LIBRARY OF CONGRESS HAS CATALOGED THE HARDCOVER EDITION AS FOLLOWS:
Gallenkamp, Charles.
Dragon hunter : Roy Chapman Andrews and the central Asiatic expeditions /
Charles Gallenkamp ; foreword by Michael Novacek ; with the cooperation
of the American Museum of Natural History.
p.   cm.
Includes bibliographical references (p.).
ISBN 0-670-89093-6 (hc.)
ISBN 0 14 20.0076 0 (pbk.)
1. Andrews, Roy Chapman, 1884–1960.   2. Naturalists—United States—
Biography.   3. Central Asiatic Expeditions (1921–1930)   I. Title.
QH31.A55 G36 2001
508'.092—dc21   [B]   00-052755

Printed in the United States of America
Set in Electra
Designed by Francesca Belanger

Contents

Foreword

In photographs taken on his famous expeditions, Roy Chapman Andrews looks exactly what an explorer should look like. We see a trim man with a well-chiseled face, clad in military khaki, sporting a Colt .38 revolver, standing in a remote, arid wasteland, and peering intensely toward some mysterious point on a far horizon. The image is arresting, larger than life, and even a bit calculated. It is easy to see why Andrews has been often suspected as the inspiration for Indiana Jones and other fictional heroes. But Andrews was of course the real thing—a remarkable explorer of great accomplishment. In the 1920s, Andrews and his team from the American Museum of Natural History struck out for the vast emptiness of the Gobi Desert in Central Asia. For the first time, a land expedition was carried out with an odd combination of primitive automobiles and a caravan of over one hundred camels to haul supplies, including the gasoline for the vehicles. It was an original, albeit, some claimed, a flamboyantly crazy scheme. To add to the challenge, the Gobi was a blank on the world map; only ancient caravan trails separated by hundreds of miles offered any semblance of navigable routes across the desert. Moreover, Andrews' goal was not simply to cross the desert, but to thoroughly explore its mountain ranges, canyons, and dune fields. This unprecedented scouring of the Gobi would bring with it extraordinary hardships. In the five Central Asiatic Expeditions carried out between 1922 and 1930, Andrews and his team endured blistering heat, icy blizzards, sandstorms, snakes, flash floods, marauding bandits, civil war, political intrigue, thievery, bribery, public relations disasters, and infuriating policy changes leveled by unstable governments in Mongolia and China.

These epic struggles should not obscure the fact that Andrews was an explorer with a serious, scholarly mission. Funding was raised and the expeditions launched with the compelling prediction that Central Asia might contain fossils of early humans that would reveal the true emergence of our own species. Ironically, that central goal was never attained; nary a scrap of genuinely ancient human bone was ever retrieved by the Central Asiatic Expeditions. Yet Andrews and his team still triumphed on the scientific front. On the track of prehistoric humans they stumbled into fossil sites with the exquisitely preserved remains of dinosaurs, ancient mammals, and other prehistoric life. The team also assembled an extraordinary collection of extant plants and animals, and conducted exhaustive surveys of the geology and geography of the Gobi. One of the most important moments in the history of paleontology—the science devoted to the study of fossils—came with the discovery of nests with dinosaur eggs in an isolated escarpment of red sandstone, evocatively dubbed the Flaming Cliffs by Andrews. The dinosaur eggs and the legendary dinosaurs, like the shield-headed *Protoceratops* and the vicious *Velociraptor*, extracted from the Flaming Cliffs, shocked and excited the whole world. For generations to follow, including the few of us who eventually found ourselves working as paleontologists, those glorious fossils at the Flaming Cliffs were the stuff of dreams and inspiration. They affirmed that a huge, mysterious world still contained things that were utterly astounding and awaiting discovery.

Such discoveries often result from serendipity as well as experience and effort. The American Museum team was actually lost on a featureless plain in the northern Gobi the day they discovered the Flaming Cliffs. While Andrews asked directions from some locals, the expedition photographer, J. B. Shackelford, took a stroll and nearly fell into the incandescent canyon, which, from Shackelford's southern vantage point, was nothing more than a thin red line at the rim of the plain. I have driven that route more than forty times in the last decade, and can attest that the cliffs are still easy to miss without a well-marked road or a GPS receiver. One wonders how paleontology might have stalled if Andrews had not been lost and had not stopped at that arbitrary spot on that fateful September day in 1922. As the title of his autobiography, *Under a Lucky Star* asserts, Andrews himself acknowledged the pervasive impact of luck in his wildly successful explorations. Yet his unique skills for leadership and organization certainly increased the odds for success. He assembled a team of experts, including the outstanding paleontologist

Walter Granger, with far more impressive scientific credentials than his own. He knew the chances for great discovery improved with the overall strength of the team, the excruciating attention to preparation and logistics, and the promotion of the enterprise.

Indeed, Andrews spent a considerable amount of time promoting the Central Asiatic Expeditions, and, simultaneously, promoting himself. This last obsession did not go unnoticed; Andrews certainly had his detractors and enemies. He was not a monolithically unblemished hero, but an interesting mosaic. He was only a mildly accomplished scientist who was responsible for some of the most important paleontological discoveries of the twentieth century. He could endure the heat and the dust of the desert, but was easily seduced by the sumptuous luxuries of high society in both New York and China's bustling capital of Beijing. He was a dedicated expedition leader, but an indifferent museum director. Admirable qualities of resourcefulness, bravery, and compassion were mixed with tendencies toward vanity, elitism, hyperbole, and unbridled ambition. And he was in the end more human and more fascinating than the comic book explorers modeled after him.

We know much of this from the hero's own tales. Andrews wrote extensively about his explorations and his adventures in numerous books and popular publications. He was a gifted storyteller, and his stories riveted many of us from childhood with their unforgettable images—camel caravans, bullet-belted bandits, and big dinosaur skeletons. There are also biographical sketches of Andrews in both children's and adult books about dinosaur exploration. Yet neither these works nor Andrews' own reminiscences flesh out key aspects of his life, his formative years as a boy romping through the woods of Wisconsin, his willingness to take on the most prosaic tasks in his early days at the American Museum of Natural History, his wanderings in Japan and China that set the stage for his full-blown scientific expeditions, his role in espionage for the United States, his entrapment in the intricacies of Asian politics, his failed first marriage, and his troubled later years coping with both his retirement from exploration and his new role as the Museum's director. This first comprehensive biography by Charles Gallenkamp superbly provides those missing portraits along with vivid and well-documented accounts of the famous exploits that led to Andrews' greatest discoveries. We see more clearly than in any other source how a passionate single-minded man converted his mad dream of unlocking the secrets of Central Asia into a triumphant scientific quest. Moreover, Gallenkamp shows us for the first time how much this high adventure was interwoven with some of the pivotal events of modern history, including the wars that led to the dominance of communism in Mongolia and China. Gallenkamp provides not only the definitive account of the Andrews saga, but also an illuminating picture of the golden age of exploration in a world much different from our own. As the author describes, Andrews' expeditions were not abandoned by lack of discoveries or hardships in the wilderness, but by the political upheavals that sealed out westerners from Central Asia for more than six decades. Today, scientists throughout the world, I among them, have been fortunate enough to gain political entry and to report that the Gobi still contains unexplored canyons and extraordinary fossils. We owe this opportunity not only to the sweeping tides of global politics. The continuing thrill and success of the fossil hunt in the Gobi, arguably the richest fossil territory in the world for the late age of the dinosaurs, is due to the man who stared intensely toward the distant horizon of a vast, uncharted desert and beckoned us all to follow him.

Michael J. Novacek
American Museum of Natural History

Preface

On the morning of April 21, 1922, three automobiles and two trucks transporting a group of heavily armed scientists, their Chinese and Mongolian assistants, and piles of equipment set out from the trading center of Kalgan, 145 miles northwest of Peking. The motorcade—a strangely incongruous sight as it roared past ancient mud-walled villages, crowds of curious peasants, and processions of camels and oxcarts—made its way westward along a narrow, jolting road to the town of Wanchuan, then passed through a gate in the Great Wall. Here the travelers turned north toward Outer Mongolia and the desolate expanse of the Gobi Desert, one of the earth's least-known areas and the object of an extraordinary journey that would carry them into the heart of this vast, almost inaccessible part of central Asia.

Led by the celebrated explorer Roy Chapman Andrews, they had embarked upon an ambitious undertaking that involved a number of untested theories, physical dangers, and political upheavals in a region generally regarded by scientists as a *terra incognita* that offered little of interest to naturalists. Undaunted by these uncertainties, the "dragon hunters," as the Chinese (who believed fossils were dragons' bones) called the explorers, proceeded to make scientific history.

Operating under the auspices of the American Museum of Natural History in New York City, and sponsored by a host of Wall Street financiers, the project—officially known as the Central Asiatic Expeditions— eventually expanded into a series of five journeys to Outer and Inner Mongolia carried out between 1922 and 1930. Andrew's daring venture was the largest, best-equipped, and most costly enterprise of its type ever launched from the United States up to that time. In concept and scope, the expeditions were unprecedented: their organization, logistics, and achievements defy the imagination. Indeed, the *New York Times'* science editor, John Noble Wilford, has called the undertaking "the most celebrated fossil-hunting expedition of this century."

It was, however, far more than a quest for ancient life in an uncharted corner of the earth. The Central Asiatic Expeditions constituted one of the most enthralling and widely publicized adventures in the annals of scientific discovery, an event that forever changed the nature of exploration. Moreover, in an age that still idolized daredevils, soldiers of fortune, and adventurers of every ilk, Andrews' role as organizer and leader of the Central Asiatic Expeditions—together with earlier escapades in Japan, Korea, China, and Mongolia—brought him extraordinary fame, making him an icon for millions of admirers and forever linking his name with the Gobi Desert's compelling mystique.

In delving into Andrews' life, I deliberately have avoided attempts at unfounded psychological analysis. As the quintessential "man of action," he devoted little time to pondering deeply philosophical or intellectual questions. Nor did he expose much of his "inner self." His publications tell us frustratingly little about his youth or family background, and his correspondence and journals only rarely lapsed into personal reflections. No doubt this was at least partly an effort to shield certain private aspects of his life against the outpouring of public attention inspired by his remarkable exploits.

Yet despite bouts of despair and frustration, Andrews was by nature optimistic, gregarious, and outgoing. His sense of humor was legendary and his enthusiasm irrepressible, though underneath these traits lurked a steely resolve that contributed to his swashbuckling, larger-than-life image. A master of diplomacy and subtle persuasion in organizing and carrying out his explorations, he was just as adept, if necessary, at forceful coercion, subterfuge, or the use of firepower. Moreover, Andrews' world was a kaleidoscope of glaring contrasts: he was equally at home in the drawing rooms of Wall Street millionaires or the salons of Park Avenue socialites, or sailing with hard-bitten whalers, languishing in a Japanese brothel, or camping with Mongol nomads.

Although this is the first adult biography of Andrews, several books for young readers about his adventures have appeared since 1930, and a recent resurgence of interest in Andrews' life has prompted numerous television documentaries and magazine articles dealing with the subject.

All too often these sources have been flawed by factual errors and misconceptions, which I have attempted to set straight for the record.

More than ten years of research have gone into this project, which carried me from the archives of the American Museum of Natural History in New York (my primary resource) to Washington, D.C.; Beloit, Wisconsin; Colebrook, Connecticut; Los Angeles, Chicago, Carmel, Dallas, Santa Fe, and ultimately to China and the Mongolian People's Republic to retrace Andrews' footsteps in the Gobi.

Regrettably, I was unable to gain access to an extensive collection of the papers of Walter Granger. As a result, it was impossible to examine material that might have added extra dimensions to this remarkable man, who not only was Andrews' close friend but also served as second in command of the Central Asiatic Expeditions, as well as chief paleontologist and the project's overall scientific coordinator. It should also be noted that many individuals associated with the Central Asiatic Expeditions did not keep journals other than scientific field notes, and their correspondence to relatives and friends has become hopelessly scattered or lost.

Of the many people and institutions that helped make this book possible, I am particularly indebted to the School of American Research in Santa Fe and its president, Douglas W. Schwartz, whose enthusiasm for the project was a constant source of encouragement. My appointment as a research associate of this institution enabled me to secure funding from individuals and foundations, administered by the School of American Research, that helped underwrite a substantial part of this undertaking. I am deeply grateful to the Ludwig Vogelstein Foundation and two of its former officers, Marga Franck and Douglas Blair Turnbaugh, for a grant in support of my initial research for this book. Additional sponsors included Stephen Goodyear; Robert N. Enfield; and the Old Taos Trail Company (all of Santa Fe); my friends Frank C. Stuart of Miami and Roberta Bishop of Boulder; William L. Kemper, Jr. and Sarah Seline, who arranged a grant from the Edward Ewing Barrow Foundation in Houston; the Earhart Foundation in Ann Arbor; the John Kittredge Anson Educational Fund; and the Carnegie Fund for Authors.

I owe a special debt of gratitude to Claire Phillips and Barbara Van Cleve of Santa Fe for their friendship and help during a particularly difficult time; and to one of my staunchest supporters, John G. Bourne, who not only provided major financial assistance to complete the project, but also invited me to make use of a secluded studio adjacent to his home in Santa Fe, where much of the manuscript was written under ideal circumstances.

One of the most rewarding pleasures I have experienced since the inception of this book has been the encouragement and cooperation extended by Andrews' son, George Borup Andrews, and his charming wife, Mary Nancy. Apart from a working relationship that afforded a wealth of insight into the personal and professional life of George's father, my wife and I have enjoyed a warm friendship with these two exceptional people. We have fond memories, too, of numerous meetings with Andrews' second wife Wilhelmina (known to everyone as Billie) and her delightful husband, Robert A. Street. Sadly, both of them recently passed away, but we relish the hours spent in their home in Carmel, talking and poring over scrapbooks of Billie's life with Andrews. Adding to the poignancy of these visits, we sat amid the priceless collection of Chinese art that once graced Andrews' palatial house in Beijing, which also served as headquarters for the Central Asiatic Expeditions.

Because this project is a joint venture with the American Museum of Natural History, the unstinting cooperation of various members of its staff was invaluable. First, I want to thank Nina J. Root, formerly the director of library services and now director emeritus. Her extraordinary knowledge of the massive archives relating to Roy Chapman Andrews and the Central Asiatic Expeditions, together with her generosity in allowing my wife and me to photocopy essential documents and her outgoing friendship—always tinged with humor—made our weeks of work in the Museum's library a rare treat. Special collections manager Joel Sweimler, now the Museum's exhibition coordinator, also performed countless favors on my behalf, especially in extracting obscure items from the depths of the archives, as did the very capable Andrea La Sala; Barbara Mathé, now the senior special collections librarian; Matt Pavlick; and Mark Katzman.

I am deeply grateful to Michael J. Novacek, senior vice president and provost of science at the Museum; Mark A. Norell, curator and division chairman of vertebrate paleontology; and Malcolm C. McKenna, Frick Curator of Vertebrate Paleontology, for their constant encouragement, guidance, and for reading portions of the manuscript for technical accuracy. (I will always remember a day spent with Mike, Mark, and their fellow scientists exploring the Flaming Cliffs deep in the Gobi, the site of many of Andrews' most famous discoveries.) And without the help of Ruth Sternfeld, I would never have navigated the Department of Paleontology's archives.

My thanks to Ellen V. Futter, president of the Museum, and her assistant Linda Cahill, for supporting this project; to Douglas J. Preston, formerly manager of publications at the Museum and presently an author living in Santa Fe, for his years of support and useful suggestions pertaining to almost every aspect of the book; the late Thomas D. Nicholson, who directed the Museum during the early stages of my work; his successor, William J. Moynihan, who was equally supportive; David D. Ryus III, who, before his departure as vice president of the Museum, first approved my proposal for the book; and L. Thomas Kelly, until recently the vice president for publication. My appreciation is also due to the staff of the Museum's Discovery Tours program, especially Penelope Bodry-Sanders, for making it possible for my wife and me to visit China and Mongolia in 1995; and to Maron L. Waxman, the director of special publishing, for her generous assistance with a wide range of issues involving marketing and promotion.

Among other persons who helped along the way, I must single out Edwin H. Colbert, one of the legendary figures in paleontology, who spent an afternoon with my wife and me at Ghost Ranch in northern New Mexico—a dinosaur hunter's paradise—reminiscing about his tenure as chairman of the Museum's Department of Vertebrate Paleontology and his memories of Andrews; Charles M. Berkey of San Antonio, Texas, whose father served as chief geologist for the Central Asiatic Expeditions; George Simson of the Center for Biographical Research at the University of Hawaii at Mānoa, who supplied me with important documents obtained from the Mongolian National Archives in Ulan Bator; the late Lowell Thomas; George E. Duck of Albuquerque, New Mexico, for making available his research into the life of the Arctic explorer George Borup, his sister Yvette Borup—Andrews' first wife—and their family background; the late Mrs. Clifford Pope of Escondido, California, who allowed me to examine her husband's correspondence from China while acting as herpetologist with Andrews' expeditions; Clive Evan Coy of Drumheller in Alberta, Canada, for sharing invaluable information gathered while compiling an exhaustive annotated bibliography of Andrews' publications; William Haskell, the present owner of Andrews' country home, Pondwood Farm, in Colebrook, Connecticut, who graciously allowed me to visit this lovely spot on two occasions; Ann Bausum of Beloit, Wisconsin; Paul Kerr, director of the Beloit Historical Society; Robert Irrmann, archivist for Beloit College; Lynn Hale of the Department of Public Relations at Lucasfilm in San Rafael, California; and Kenneth and Katherine Ferguson, my wife's children, who willingly relinquished their mother's company for long periods while she labored with me on the manuscript. Twice, Kenneth kept our research on schedule with loans, a true measure of his faith in our obsession.

To my agent, Owen Laster of the William Morris Agency, I express my heartfelt thanks for his expert guidance; to Kathryn Court, president and publisher of Penguin Books, for her sustained interest in the subject over the years; to my editor, Michael S. Millman, for his belief in the project and patience during the process of polishing the manuscript; Francesca Belanger for the book's design; Paul Buckley and Jesse Reyes of Viking's art department for a superb jacket; Barbara Campo for production editing; Beth Greenfeld for her skilled copy editing; and Carolyn Coleburn and Paul Slovak for their creative direction of the book's publicity and promotion.

I owe a profound debt of gratitude to my late parents, Norma Benton and Charles O. Gallenkamp, for first awakening me to the thrill of adventure in far-off places and encouraging me to follow its lure through books and travel. And I regret that I will never be able to convey my personal appreciation to Roy Chapman Andrews, whom I met on several occasions during his years as director emeritus of the American Museum of Natural History, where I was working in the afternoons as a volunteer apprentice in the late 1940s. I will always remember the day he saw me buried in a book in what was then the fifth-floor reading room. It happened to be his own summary account of the Central Asiatic Expeditions, *The New Conquest of Central Asia,* and Andrews took the time to point out specific sections he thought I would find especially exciting. How right he was! Nor did he know that episode would awaken a lifelong interest in his explorations.

Finally, my wife, Karen L. Wright, has been a constant source of inspiration and a tireless partner throughout the writing of this book. I cannot praise too highly her efforts in sorting out enormous quantities of research material, deciphering Andrews' difficult handwriting in his correspondence and journals, coping with daunting organizational problems—including researching and writing several particularly complicated chapters, numbers 18, 19, and portions of 20—editing the manuscript, and contributing a series of lucid drawings illustrating some of Andrews' most significant paleontological discoveries. Her irrepressible spirit permeates every aspect of this book; by rights, her name should appear on the jacket as my collaborator.

Charles Gallenkamp
Santa Fe, New Mexico

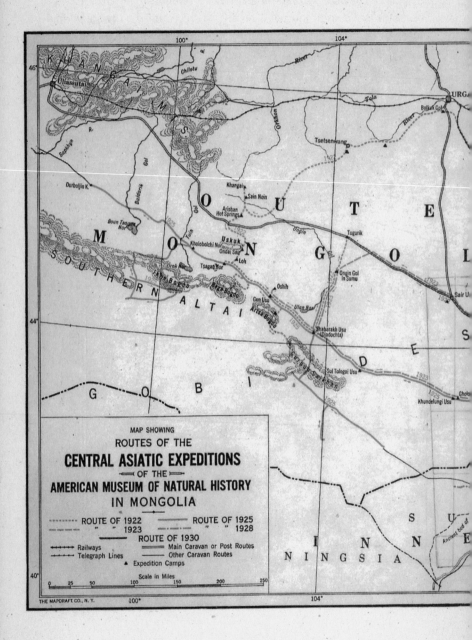

MAP SHOWING
ROUTES OF THE
CENTRAL ASIATIC EXPEDITIONS
OF THE
AMERICAN MUSEUM OF NATURAL HISTORY
IN MONGOLIA

ROUTE OF 1922   ROUTE OF 1925
"   " 1923   "   " 1928
ROUTE OF 1930
Railways   Main Caravan or Post Routes
Telegraph Lines   Other Caravan Routes
▲ Expedition Camps

Scale in Miles
0   25   50   100   150   200   250

THE MAPDRAFT CO., N.Y.

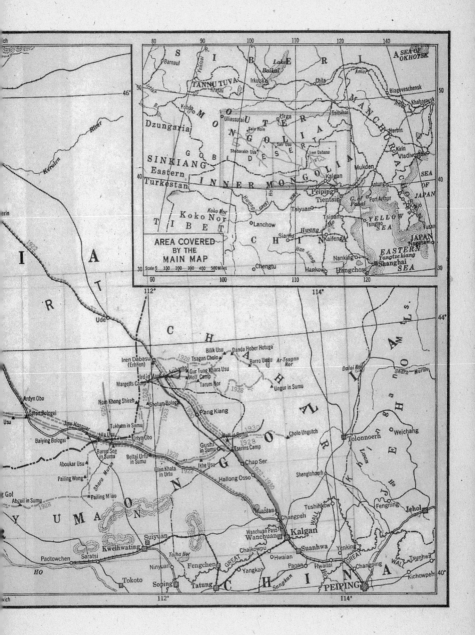

A Note on Place-Names

In dealing with geographical locations in China and Mongolia, one encounters bewildering variations in spelling. Virtually all of the major cities mentioned in this book have undergone name changes, often more than once, since Roy Chapman Andrews wrote about them. Peking is now Beijing; Kalgan has become Ch'ang-Chai-k-'ou or Changchiak'ou; Tientsin was changed to Tienching and Tianjin; Nanking appears on many maps as Nanjing; Iren Dabasu in Inner Mongolia became Erhlien or Erlin. In the Mongolian People's Republic—formerly Outer Mongolia—its capital, Urga, has been known since 1924 as Ulan Bator, Uaalanbaatar, and Ulaan Baatar. Alternate spellings of other cities and geographic features are noted in the text.

To conform with the primary sources cited in my research, I have adhered to the spelling most often used by Andrews, although even he was not always consistent in his spelling of various locations. As for the names of numerous places at which the Central Asiatic Expeditions worked or camped—such sites as Ondai Sair, Ula Usu, Wild Ass Camp, Sair Usu, Ardyn Obo, Pailing Miao, Wolf Camp, Koko Nor, Viper Camp, Gosho-in Sumu, and many others—these were nomad encampments, wells along caravan trails, religious shrines, or names given to camps by the explorers. Nearly all of them have vanished from modern maps, and only a few can be identified with certainty. Unfortunately, most of the lamaseries, missions, and temples mentioned by Andrews were either destroyed under Mongolia's Communist regime or allowed to fall to ruin.

Scientific and Technical Staff
of the Central Asiatic Expeditions

Roy Chapman Andrews—Leader and Zoologist, 1921–1930
T. Badmajapoff—Mongolian Political Representative, 1922
Radcliffe H. Beckwith—Geologist, 1926–1927
Charles P. Berkey—Chief Geologist, 1922–1925
F. B. Butler—Assistant Topographer, 1925
Ralph W. Chaney—Paleobotanist, 1925
S. Bayard Colgate—Chief, Motor Transport, 1922
Joel Eriksson—Agent in Mongolia, 1926–1930
A. Z. Garber—Surgeon, 1930
A. W. Grabau—Research Associate, 1922–1930
Walter Granger—Chief Paleontologist, Second in Command, 1921–1930
W. P. T. Hill—Topographer, 1928
G. Horvath—Motor Transport, 1928
Albert F. Johnson—Assistant in Paleontology, 1923
C. Vance Johnson—Motor Transport, 1923
Peter C. Kaisen—Assistant in Paleontology, 1923
F. A. Larsen—Interpreter and Expedition Agent, 1922–1926
H. A. Loucks—Surgeon, 1925
Norman Lovell—Motor Transport, 1925
W. D. Matthew—Paleontologist, 1926
Frederick K. Morris—Geologist, 1922, 1923, and 1925
Nels C. Nelson—Archaeologist, 1925
George Olsen—Assistant in Paleontology, 1923–1925
J. A. Perez—Surgeon, 1928
Alonzo W. Pond—Archaeologist, 1928
Clifford H. Pope—Herpetologist, Chinese Division, 1921–1926
L. B. Roberts—Chief Topographer, 1925

H. O. Robinson—Assistant Topographer, 1925
J. B. Shackelford—Photographer, 1922, 1925, 1928
L. Erskine Spock—Geologist, 1928
Père Teilhard de Chardin—Geologist, 1930
Albert Thomson—Assistant in Paleontology, 1928–1930
W. G. Wyman—Topographer, 1930
J. McKenzie Young—Chief, Motor Transport, 1923–1930

PART ONE

Under a Lucky Star

The explorer is the poet of action
And exploration is the poetry of deeds

—Vilhjalmur Stefansson

CHAPTER 1

"I was born to be an explorer. There never was any decision to make. I couldn't do anything else and be happy. . . . The desire to see new places, to discover new facts—the curiosity of life always has been a resistless driving force in me." So Roy Chapman Andrews wrote in the foreword to his book *This Business of Exploring*. It was a conviction he repeated many times in print and from lecture platforms, and the course of his life confirmed its validity.

Andrews was born in Beloit, Wisconsin, on January 26, 1884, at approximately two o'clock in the morning. Upon first seeing his son, Andrews' father noticed what he perceived to be an Oriental cast to the infant's eyes, causing him to jokingly declare, "Why I've begot a Chinaman!" It was a prophetic remark in view of the years Andrews lived in China and the impact of the Orient upon his career.

Andrews grew up in a decidedly conventional atmosphere. His father, Charles Ezra Andrews, was a native of Worthington, Indiana, who had moved to Beloit in 1873, lured by the town's reputation as a burgeoning industrial center. Within a few years, he was earning a comfortable living as a wholesale druggist whose hard work and involvement in civic affairs made him widely respected in the community.

Some time in 1880, Andrews became engaged to Cora May Chapman, the daughter of an entrepreneur named James A. Chapman. Originally from Utica, New York, he had moved to Beloit with his wife and children in 1858 and quickly established a successful real estate and insurance business. Although Cora was two years old when her family arrived in Beloit, she always considered herself a "native." After graduating from high school, she continued to live with her family, devoting much of her energy to community organizations. By the time of her engagement to Andrews, she had grown into a mild-mannered, studious woman who was fond of reading fiction and books on history and travel.

On December 8, 1880, Charles and Cora Andrews were married in a small Protestant ceremony held in the Chapmans' home, and they quickly settled into a pleasant if somewhat staid existence. With their two children—a daughter named Ethelyn May was born in 1882, followed by Roy's arrival two years later—the couple enjoyed a harmonious marriage that lasted until Cora's death in 1935. While Charles never became wealthy as a wholesale drug salesman, he managed to provide his family with a comfortable house in one of Beloit's better neighborhoods, an above-average share of everyday amenities, and a cabin in the woods west of town, which he built for weekend outings.

Like most Beloiters, their lives reflected a sober legacy of initiative, self-reliance, and devotion to conventional pieties derived from the town's deeply entrenched New England roots. Nearly all of Beloit's founders had come from New York, Connecticut, Massachusetts, Vermont, New Hampshire, and Maine at the instigation of the New England Emigrating Company, organized in Colebrook, New Hampshire, in 1836. Its sole objective was to recruit settlers from northeastern states for relocation in Beloit, where the firm had invested heavily in the area's economic development.

At the time of Roy Chapman Andrews' birth in 1884, the population of Beloit numbered about six thousand people. Situated on high bluffs along both sides of the Rock River, the town was already a thriving manufacturing center. By the 1890s, over two dozen factories were producing windmills, pumps, waterwheels, bicycles, papermaking machines, scales, steam and combustion engines, woodworking equipment, plows, and various other items that were being marketed throughout the United States and exported worldwide. Apart from sprawling industrial areas, Beloit contained a pleasing mélange of tree-lined residential neighborhoods, well-manicured parks, and a flourishing business district. Its buildings—many of them constructed of limestone from quarries along the river—embodied a mixture of Greek Revival, Victorian, and Gothic architecture. And compared to most midwestern towns of similar size, Beloit offered what residents lauded as a "refined and cultivated atmosphere." Its cultural diversions boasted choral groups, bands, theatrical companies, a Philharmonic Society "devoted to pleasure and appreciation of good music," a reading and poetry club, an elegantly appointed opera house, occasional art exhibitions, lectures, and appearances by internationally famous musicians, actors, and dancers.

Beloit was a pleasant-enough place in which to grow up, and Andrews' memories of his boyhood there were tinged with nostalgia, especially for the nearby river and woodlands, which afforded an ideal training ground for his fledgling interest in natural history. Yet there were the usual disadvantages of a small town, and for someone possessed of Andrews' curiosity, restlessness, and thirst for adventure, his horizons would inevitably expand beyond anything Beloit could offer him as an adult. Above everything loomed the certainty that his dream of becoming an explorer and naturalist could only be attained, if at all, in more cosmopolitan surroundings: cities with active, well-financed museums or other scientific institutions. It was this reality that would cause Andrews to leave Beloit immediately after graduating from college, returning only for brief visits with his parents.

Unfortunately, Andrews' books, private papers, and correspondence convey only a fragmented portrait of his family background and early years in Beloit. In his autobiography, *Under a Lucky Star* (1943), he devoted barely ten pages to his first twenty-one years, explaining that no one but himself would give "a tinker's damn" about his boyhood. When referring to his ancestry, he was equally terse, dismissing the topic with a remark he once heard that "people who are too much concerned with the pedigree of their forefathers are apt to be like potatoes—the best part of them is underground."

Andrews' writings contain no mention of relatives other than brief references to his mother and father. A single exception occurs in his book *An Explorer Comes Home* (1947), a sequel to his autobiography, in which he traces in less than a paragraph the lineage of his great-grandfather, Noah Andrews, back to one John Andrus, who came to America from Essex County, England, in 1640.

Andrews enjoyed a comfortable middle-class upbringing. He was raised in a secure, supportive household, and his relationship with his parents was always quite close. Until his graduation from college, he lived with his family in the two-story, fourteen-room house at 419 St. Lawrence

Avenue that his maternal grandfather, James A. Chapman, had purchased in 1859. Gregarious and fun-loving by nature, Andrews possessed a ready sense of humor and a fondness for pranks. Even at an early age, he was fiercely independent, strong-willed, and high-strung. He prided himself on financial self-reliance, and to earn extra money, he worked at odd jobs such as gardening, taking care of horses, distributing circulars, and driving a delivery wagon for a bakery. Although he was thin and not especially muscular, Andrews was athletic and unusually strong, and his boundless energy often caused his parents to issue vain warnings against exhausting himself.

Yet there was a reclusive side to his personality as well, an "inner life," jealously guarded from intrusion, that centered around his love of reading, nature lore, and solitary days spent in the countryside hunting, studying wildlife, and camping. Photographs of Andrews taken at the ages of eight, fourteen, and twenty show a handsome youth with penetrating eyes, a high forehead, and a resolute jaw. Older-looking than his years, he was once described by a family friend as "already having the appearance of someone destined to be famous."

Andrews received most of his secondary education in public schools. When he was sixteen, he transferred for a year to the Beloit College Academy, a private institution established to serve as a "feeder" for the college itself, since few graduates of the local high school could, one observer noted, "hurdle the high wall of classical requirements for entrance [primarily Greek, Latin, and rhetoric] by which the college was surrounded." Never an exceptional student, Andrews somehow managed to navigate the academy's rigorous courses with satisfactory grades, though all forms of mathematics were agonizing to him, remaining what he called his *"bête noire"* throughout life. "I add correctly," he confessed in his autobiography, "only as high as I can count on my fingers, and subtract and divide very uncertainly."

In September 1902, Andrews enrolled at Beloit College. Notwithstanding the devout New England puritanism of its founders, the college had always been remarkably progressive. Beloit's faculty adopted a liberal approach to education that encouraged the pursuit of "not only Bible truth but all truth." As an outgrowth of this ambiguous mandate, the school offered a nonsectarian curriculum that emphasized humanities and science, along with mandatory biblical studies. Even Darwin's theory of evolution, widely shunned in many universities, was being taught at Beloit by the 1890s due to the influence of an enlightened professor named George L. Collie, a Harvard-trained scientist who became the first curator of the Logan Museum of Anthropology, which opened on the campus in 1894.

During his four years in college, Andrews worked just hard enough to earn acceptable grades, and his transcript reveals a student who was capable but erratic. He received above-average marks in English, science, economics, German, archaeology, history, and rhetoric—with a scattering of A's—though the rest of his grades were mediocre to poor. But regardless of his lack of academic zeal, Andrews' involvement in extracurricular activities was another matter: he played on his class baseball team, served as secretary-treasurer of the Boating Club, joined the Sigma Chi fraternity, and maintained a busy social life with the opposite sex, which caused him, he lamented, to "[waste] a lot of time."

On one occasion while attending Beloit, Andrews demonstrated the persuasive charm that would see him through many tight spots in later life. Knowing he would never pass freshman mathematics, a requirement for graduation, he turned his attention to his English teacher, an attractive young woman who happened to be in love with the mathematics professor. Andrews excelled in English, and intensifying his efforts even more, he was awarded the highest marks in literature and composition, along with frequent invitations to his English teacher's home for tea. One day he mentioned, not very subtly, that he knew he was unable to pass the dreaded mathematics course and would be dropped from Beloit at the semester's end, causing her to lose her star English pupil. On the other hand, Andrews hinted, "If she could persuade her suitor to give me a passing mark in mathematics the situation would be saved. She did and it was!"

Unquestionably, Andrews' real interests lay outside the classroom, and formal education was of secondary importance in shaping his intellectual development. It did little to nurture his deeply rooted determination to become a naturalist and explorer—ambitions that dominated his life from an early age and led him down paths of independent study. As a child, he adored stories involving epic journeys, wild animals, adventure, and scientific discovery. His favorite book was *Robinson Crusoe*, which his mother read to him a dozen times. By the age of eight, he showed a consuming interest in nature, and began to spend most of his spare time wandering in the woods or along the banks of the Rock River. Equipped with binoculars and a notebook, he delighted in observing wildlife and recording the migrations of birds. Occasionally, he visited the Field Museum of Natural History in Chicago, ninety miles away, spending hours studying its exhibits or delving into its library. He also devoted painstaking effort to assembling a "private museum" of his own—a collection of minerals, fossils, stuffed animals, insects, bird skins, Indian artifacts, and dried plants, which were carefully labeled and displayed in the attic of his house. His love of the wilderness led to a passion for hiking and camping trips, during which he liked to test his ability to live off the land, eating mostly berries, fish, and game. "I was like a rabbit," he wrote, "happy only when I could run out of doors. . . . Whatever the weather, in sun or rain, calm or storm, day or night, I was outside, unless my parents almost literally locked me in."

For Andrews' ninth birthday, his father, who encouraged his son's sporting instincts, gave him a single-barrel shotgun with which he learned to hunt. In the process, he once mistakenly blew up three "geese," unaware that they were actually pneumatic decoys belonging to another hunter. "At the roar of my gun," Andrews recounted, the fragile geese "slowly collapsed with a gentle hissing sound." The incident so amused his father, who sorely disliked the decoys' owner, that he promptly bought Andrews a more-powerful double-barrel shotgun.

Using William T. Hornaday's book, *Taxidermy and Home Decoration,* Andrews taught himself to mount animals and birds. He soon became so skilled that he acquired a license from the Wisconsin Conservation Department and started a part-time business mounting trophies for hunters, the proceeds from which paid for most of his college tuition. Andrews exhibited his handiwork in stores, barbershops, and saloons; and his talents were solicited by family friends and schoolmates who brought him their dead pets to be mounted—everything from dogs, cats, and parrots to snakes, lizards, and turtles.

Throughout his formative years, Andrews devoured books on natural history, travel, and exploration. He carried a tattered copy of Frank M. Chapman's *Handbook of Birds of Eastern North America*—his "Bible," as he called it—on every field trip. He pored over narratives of the great nineteenth-century African explorers: Richard Burton, John Hanning Speke, Samuel Baker, David Livingston, and particularly Henry M. Stanley, whose *In Darkest Africa* so thrilled Andrews that he "could scarcely sleep at night after reading it." In a secondhand bookstore, he bought a copy of Charles M. Doughty's classic *Travels in Arabia Deserta*, which may have helped awaken his attraction to deserts. He eagerly followed accounts of Fridtjof Nansen's daring polar voyage in 1893–1896 aboard his specially built ship, the *Fram*; and he was enthralled by Sven Hedin's two-volume work, *Central Asia and Tibet: Toward the Holy City of Lassa*, an English translation of which appeared in 1903.

Added to Andrews' interest in nature, hunting, and taxidermy, these books intensified his enthusiasm for science and exploration to the point of obsession. "From the time that I can remember anything," he declared, "I always intended to be an explorer, to work in a natural history museum. . . . Actually, I never had a choice of profession. I wanted to be an explorer and naturalist so passionately that anything else as a life work just never entered my mind." Exactly how he would achieve these ambitions was yet unclear, though at that point, he wrote, "I never let practical considerations clutter my youthful dreams."

While he was still in college, two unforeseen events gave Andrews a decisive push toward his goals. Midway through his junior year, he experienced a tragedy that profoundly affected his outlook. His closest friend was a twenty-three-year-old New Yorker named Montague White, an instructor in rhetoric at Beloit College. On March 31, 1905, while hunting ducks near the junction of Young's Creek and the Rock River, about seven miles north of Beloit, a canoe in which Andrews and White were trying to cross the river suddenly capsized. White, an excellent swimmer, quickly vanished in the icy, turbulent water, swollen by melting snow and a week of heavy rains. A moment later, he surfaced briefly, then disappeared again. Andrews, meanwhile, was swept downstream toward a partially submerged clump of willows. By clinging to branches, he was able to pull himself to safety and crawl more than half a mile to a farmhouse, tearing a deep gash in his leg on a barbed wire fence. Suffering from cold and exhaustion, he was put to bed by the farm's owners, who summoned the police. At six o'clock that evening, a search party recovered Montague White's body; seized by cramps, he had drowned only a few yards from where he first went under, within easy reach of the riverbank.

Devastated by his friend's death, Andrews lapsed into severe physical and emotional trauma. He lost forty pounds during the next few weeks, and the slightest excitement caused him to tremble violently. For almost six months he remained withdrawn, spending as much time as possible wandering in the woods "with field glasses and notebook studying birds," he wrote, "or sleeping in the sun." Though he eventually recovered from the accident's immediate effects, it may have been responsible for the onset of a slight nervous tic in his left cheek, which persisted for the rest of his life and was most noticeable when he became agitated. Andrews' near brush with death also became an agonizing catalyst that awakened a fierce determination to pursue his long-cherished plans. "Monty's drowning," he declared, "overwhelmed me with the frailty of my own existence. I realized as never before the importance of time, the need to focus my energy, and the necessity of exploiting every opportunity if I ever expected to fulfill my ambitions."

Less than a year later, just such an opportunity presented itself. Edmund Otis Hovey, assistant curator of geology at the American Museum of Natural History in New York, came to Beloit College in 1906 to lecture. As an authority on volcanoes, his subject was the eruption of Mount Pelée, which four years earlier had killed thirty thousand people on the island of Martinique in the West Indies. Ever since Andrews could remember, the American Museum had represented the pinnacle of his aspirations. He had read every magazine article he could find on its expeditions to far-off places, and during his visits to the Field Museum in Chicago, he never failed to scour its library for back issues of the *American Museum Journal*, the forerunner to *Natural History* magazine. Long before Edmund Hovey's lecture in Beloit, Andrews had extolled the American Museum as "the one place I would most want to work if given a choice."

Eager to discuss the possibility of employment with a member of that august institution's staff, Andrews hounded Hovey's hotel until he cornered the courtly, rather shy geologist in the lobby. Hovey, who proved to be "exceptionally pleasant," sat patiently while Andrews expounded on his interest in exploration, natural history, and taxidermy; Hovey even agreed to walk around the corner to Moran's Saloon, where several deer heads and birds mounted by Andrews were displayed on the walls. Impressed by the would-be naturalist's abilities, Hovey offered to speak to the American Museum's director, Hermon Carey Bumpus, on his behalf. He also suggested that Andrews write to Bumpus directly about prospects for a job.

Andrews immediately composed a letter to Bumpus outlining his qualifications and requesting an interview. The reply was courteous, though hardly promising. In a brief paragraph, Bumpus conveyed his regrets that no positions were currently open, but he offered to meet with Andrews if he ever came to New York on other business, cautioning him that it would be unwise to make a special trip. Andrews remembered that he and his mother "were greatly excited at the letter," though his father, "more of a realist than either of us, made some uncalled-for remark about not counting unhatched chickens."

After receiving his bachelor of arts degree in June 1906 (a diploma he felt "had not really been earned"), he resolved to visit New York at once. Spurred on by the gnawing concern regarding his future that had plagued him since Montague White's death, Andrews declined a surprise graduation present arranged by his parents: a two-week fishing trip to northern Wisconsin. Instead, he left for New York early in July with a train ticket purchased by his father—in place of the fishing trip—and $30 he had earned mounting trophies.

On his way East, Andrews stopped in Chicago to investigate job possibilities at the Field Museum, after which he paid a visit to the Carnegie Museum in Pittsburgh, but he received no encouragement at either place. When he reached New York on the evening of July 6, his first sight of its skyline was from the Twenty-third Street ferry. "The magic city," he mused, "was more beautiful than anything of which I had dreamed. All my fears vanished. . . . I knew it was my city."

The following morning, Saturday, July 7, Andrews confronted the imposing façade of the American Museum of Natural History on Central Park West at Seventy-ninth Street. At eleven o'clock, he was ushered into the director's fifth-floor office overlooking Central Park. He introduced himself to Hermon Bumpus, a tall, lean energetic man with thinning hair and a small black mustache, who had formerly served as the Museum's curator of invertebrate zoology and taught comparative anatomy and biology at Brown University.

Bumpus listened indulgently as Andrews pleaded his case for a job. Nevertheless, all hope appeared to vanish when the director reiterated what he had said in his letter: that no suitable positions were available. But a moment later, a curious exchange occurred, one that Andrews subsequently retold many times in print. Almost unthinkingly, he blurted out that he was not asking for a *position*; he simply wanted to work at the Museum in any capacity, even if he did nothing more than clean floors. Bumpus protested that a man with a college degree would never be happy scrubbing floors. Whether by a stroke of well-calculated strategy or sheer desperation, Andrews countered that he would certainly not scrub just any floors, but the Museum's floors, he insisted, were *different*. Unable to resist this outburst of youthful ardor, Bumpus relented. He agreed to hire Andrews as an assistant in the taxidermy department at $40 a month.

After lunching with Bumpus at a nearby restaurant called the Rochelle, where Andrews vividly recalled ordering "cold salmon and green peas," he was taken back to the Museum to meet James L. Clark, a gifted young taxidermist and wildlife sculptor. In his autobiography, *Good Hunting: Fifty Years of Collecting, and Preparing Habitat Groups for the American Museum,* Clark recounted how Bumpus brought Andrews to his studio, introduced him, and announced that "he has just come here to take a job. . . . I don't know what he will do, but if you'll give him a working space and a desk somewhere, I'll find something for him."

Instructed to report to the Museum promptly at nine o'clock on Monday morning, Andrews arrived at 8:15, having stayed up "most of the previous night too nervous to think about sleeping." He crossed Central Park and sat on a large granite boulder near the entrance to Eighty-first Street. Gazing at the Museum's massive stone façade, Andrews pondered his future with a mixture of apprehension and elation. "Finally, a few minutes before nine," he wrote, "I shut my eyes and made a little prayer, then walked to the entrance on Seventy-Seventh Street and, for the first time, went through the doors of the American Museum as an employee."

CHAPTER 2

An aspiring naturalist could not have asked for a more stimulating place to work than the American Museum of Natural History. Since its founding in 1869, this now-venerable institution had rapidly grown in scope and prestige. The construction of its present complex of buildings—an architectural mosaic of Victorian Gothic and Romanesque Revival (the pseudo-Roman façade adorning its front entrance was added in 1936)—was begun in 1873 on a twenty-three-acre site bordered by Central Park West, Columbus Avenue, and Seventy-seventh and Eighty-first streets. By the time Andrews arrived in 1906, the Museum employed twenty-five scientists, plus a sizable administrative and technical staff. In addition to offices, laboratories, a library, and exhibition halls, it housed an impressive collection of birds, mammals, reptiles, fish, insects, botanical specimens, minerals, fossils, and anthropological material. Its expeditions were already penetrating some of the earth's remotest areas in search of scientific data and collections, and its exhibits drew approximately 500,000 people every year.

Having also gained prominence among New York's social elite, the Museum was frequently the scene of receptions, black-tie dinners, lectures, and fund-raising events. Important visitors from around the world—royalty, statesmen, explorers, business tycoons, and scientists—were sumptuously entertained in its boardrooms and exhibition halls. And its list of benefactors included such luminous names in financial circles as Morgan, Frick, Vanderbilt, Astor, Rockefeller, Dodge, Warburg, Baker, Phelps, Jesup, and other affluent supporters, who underwrote a vigorous program of exploration, acquisition, research, and publication.

Andrews described his first weeks at the Museum as "the most thrilling time of my life." Every morning, he began by mopping the floor in the taxidermy studio, for Bumpus had taken his plea literally; the rest of the day was usually spent helping James Clark, known to everyone as Jimmy, with the preparation and mounting of animal skins. Occasionally, Bumpus gave him special tasks involving cataloging, writing labels, or arranging exhibits, and the director sometimes dropped by the studio to inspect the floor, "to see," Andrews wrote, "if the college degree had got in the way of a mop." Any extra time he managed to find in his busy schedule, mainly evenings, weekends, and holidays, was devoted to reading books and articles on natural history in the Museum's library, studying the collections, or learning photography.

After eight weeks, Bumpus was sufficiently impressed with Andrews' work to give him a $5-a-month raise. The director was quick to recognize that his energetic protégé—now a handsome young man slightly over six feet tall, with intense blue eyes, and prematurely thinning hair—was articulate and personable. He was equally aware that Andrews possessed great charm and formidable powers of persuasion. Along with the increased salary, Bumpus showed his approval in other ways. He allowed Andrews to assist with planning new exhibits, and asked him to conduct important visitors on tours of the building. He also began introducing Andrews to senior members of the Museum's scientific staff and its officers, including the president of the board of trustees, Morris K. Jesup, a railroad builder and investment banker whose estate ultimately provided the Museum with a $6 million endowment.

Another titan within the Museum's hierarchy was Henry Fairfield Osborn, regarded by Andrews, with ample justification, as "fiercely intimidating." Heir to a railroad fortune, a renowned paleontologist and teacher, and a man of enormous influence, Osborn was to succeed Morris Jesup as the trustees' president in 1908. Awed by this millionaire scientist's exalted reputation and imperious manner, Andrews frequently lingered around a display of meteorites near the Museum's entrance at one o'clock "to get a sight of Professor Osborn when he went to luncheon." As unlikely as it may have seemed at the time, Osborn was destined to become Andrews' friend, mentor, and a major force behind his most important achievements as an explorer. Even then, as the admiring neophyte was beginning his apprenticeship at the Museum in 1906, Osborn was engrossed in paleontological research that provided the impetus for the Central Asiatic Expeditions, which Andrews would set about organizing fourteen years later.

While working together in the taxidermy studio, Andrews and Jimmy Clark developed a close friendship. In September, Andrews gave up his rented room and moved into an apartment on 143rd Street with Clark and two other roommates from the taxidermy department. Both men were almost the same age—Clark, born in October of 1883, was four months older than Andrews—and each had been "discovered" by Hermon Bumpus. In Clark's case, Bumpus recruited him in 1902, when he was eighteen years old, a part-time art student at the Rhode Island School of Design and employed by the Gorham Silver Company in Providence. Recognizing Clark's talent as a sculptor, Bumpus originally hired him to model animals, but within a few months, he was sent to Chicago to learn a new method of taxidermy perfected by the brilliant artist-inventor Carl Akeley at the Field Museum of Natural History.

Akeley had revolutionized the art of taxidermy. By incorporating sculptural techniques into his creations, he was able to discard the traditional practice of stuffing animals with straw, papier-mâché, or excelsior, which usually resulted in relatively crude replications of the original specimens. Instead, Akeley utilized portions of an animal's skeleton, supported by iron rods and mounted in the desired position, as a frame on which the muscles and tendons were modeled in clay, relying on precise specifications taken from drawings and measurements made at the time the animal was killed. A plaster impression of the sculptured form was then used to cast a mannequin over which the carefully prepared skin was fitted. With the addition of glass eyes and painted facial details, the completed specimen was durable, anatomically correct, and lifelike.

After studying with Akeley for three months, Clark had introduced these procedures at the American Museum. (Akeley later left Chicago and joined the Museum's staff in 1909.) When Andrews arrived on the scene, he soon familiarized himself with Akeley's methods by assisting Clark, though having no ability as a sculptor, he concentrated on the preparation and mounting of skins. Writing of their early days in the taxidermy studio, Clark commented "that Roy and I started out [there] to be great and good museum men: Roy becoming great—while I became good."

Clark subsequently opened a commercial taxidermy studio that catered to an impressive list of celebrity big-game hunters, including Theodore Roosevelt and George Eastman, the Kodak magnate; and for almost twenty years, he served as manager and later president of the Akeley Camera Company, which manufactured an innovative motion picture camera invented by Carl Akeley. Even though these outside ventures were highly profitable, they remained secondary to Clark's Museum-related projects. Acquiring the title of assistant director for preparation and installation, he made five expeditions to Africa and trekked through Alaska, Canada, the western United States, central Asia, and Indo-China to collect animals. He also supervised the creation of three of the Museum's most famous attractions: the Vernay-Faunthorpe Hall of South Asiatic Mammals, the Hall of North American Mammals, and the spectacular Akeley Memorial Hall of African Mammals—a tour de force in the installation of habitat groups designed to exhibit animals in settings that duplicate their natural environment, with painted backgrounds showing actual locations and using exact reproductions of trees, plants, flowers, and rocks.

Scarcely had Andrews settled into his duties at the Museum before whales entered his life, unexpectedly and with far-reaching consequences. It began when Bumpus asked him to assist Clark in constructing a life-sized model of a blue whale. Undaunted by the colossal size of the model—it was seventy-six feet in length—Andrews and Clark devised a method for fabricating the whale using papier-mâché over a framework of angle iron, wood, and wire mesh. The result was so successful that the huge blue whale, lifelike in every detail, hung for sixty years in a third-floor gallery and became one of the Museum's most popular exhibits.

In the midst of this project, Andrews found himself involved in salvaging the remains of a North American right whale that washed ashore on February 7, 1907, at Amagansett on Long Island after being killed by hunters. Seizing the opportunity to secure badly needed material for the Museum's cetacean collection, Bumpus dispatched Andrews and Clark to purchase the carcass from the whalers for the commercial value of the baleen, or whalebone (about $3,200), photograph the animal, obtain anatomical notes and measurements, and retrieve the skeleton.

Exuberant over the prospect of setting out on his first "expedition" after only seven months at the Museum, Andrews, along with Clark, reached Amagansett on February 22. The temperature stood at twenty degrees below zero, a freezing wind was blowing off the ocean, and the fifty-four-foot whale had already sunk several feet into the sand at the surf's edge. To get at the skeleton, it was necessary to cut away roughly fifty tons of flesh, a job requiring the help of local fishermen who demanded exorbitant wages to work in the bitter cold. On the second day, with only the skull and a few ribs separated from the bloody carcass, a violent storm struck the coast without warning, forcing Andrews and Clark to hastily anchor the whale with ropes attached to wooden stakes. After seventy-two hours of rain, gale-force winds, and churning tides, the carcass was completely buried; there was nothing to mark its location but the anchor ropes extending down into the sand.

By then, the fishermen refused to work any longer under those conditions—the temperature now reached a high of twelve degrees at midday—leaving Andrews and Clark to carry on alone for three days. Repeated efforts to uncover the skeleton ended in frustration, since each time a shovelful of sand was removed, the depression filled with water. Groping blindly in icy mud up to their elbows, they were finally able to disarticulate a few vertebrae with knives, constantly stopping to warm their hands over a driftwood fire. On the fourth morning, the temperature started to rise, and half a dozen men reluctantly agreed to help them, "more out of shame at watching us struggle," Andrews commented, "than the high wages." Even so, it still required another week of exhausting labor before the skeleton was rescued from its sandy grave and packed for shipment to New York.

Despite Andrews' complaint that he had "never suffered more from any experience," the episode at Amagansett, combined with his work on the blue whale replica, left him with an avid interest in Cetacea—the order of marine mammals that encompasses whales, porpoises, and dolphins. After delving into the sparse scientific literature then available on the subject, Andrews concluded that cetology offered an ideal opportunity to establish himself as a zoologist, with unlimited possibilities for original research. As a first step toward this objective, he wrote a description of the North Atlantic right whale he and Clark had collected. The resulting paper, "Notes Upon the External and Internal Anatomy of *Balena glacialis* Bonn," appeared in the Museum's *Bulletin* for 1908, marking Andrews' first foray into scientific publication.

Whales had another beneficial effect on Andrews' career: they liberated him from odd jobs in the taxidermy studio. In the spring of 1908, he was transferred to the Department of Mammalogy and Ornithology, where he came under the influence of two eminent naturalists. One was Joel A. Allen, the Museum's curator of birds and mammals, editor in chief of its zoological publications, and a noted taxonomist. His colleague in the department was Frank M. Chapman, one of the world's foremost ornithologists, founder of the magazine *Bird Lore* (later renamed *Audubon*), and a pioneer in the use of habitat groups for exhibiting wildlife. Associating on a daily basis with Chapman, whose classic *Handbook of Birds of Eastern North America* Andrews had so admired as a youth, would have been "adequate compensation," he insisted, for his elevated status at the Museum. But Andrews' new position—he was now officially an assistant in mammalogy—increased his salary to $100 a month, gave him access to the zoological collections, and allowed him to pursue his study of cetaceans with the Museum's blessings.

Early in 1906, one of the Museum's trustees, George S. Bowdoin, a wealthy partner in J. P. Morgan and Company, had been persuaded to donate $10,000 to enlarge the cetacean collection; a portion of these funds, in fact, had paid for construction of the blue whale model and recovery of the Amagansett right whale. At first, Bowdoin was reluctant to put up money for something he considered as "unglamorous" as whales, especially, Andrews noted, when his friends "were giving Rembrandts and Van Dycks to the Metropolitan Museum of Art."

But Andrews soon devised an ingenious means of capitalizing on Bowdoin's investment. He conceived an ambitious program of research that involved the collection of anatomical data at shore-whaling stations in British Columbia and Alaska. Shore whaling was still a relatively new procedure, and because whales killed at these facilities were processed out of water on wharves, rather than at sea as had traditionally been the custom, Andrews felt it would be easier to examine specimens while they were being butchered. Outlining the plan to Bumpus and Joel Allen—now his immediate superior in the Department of Mammalogy—Andrews argued convincingly that such a project would yield a wealth of information at relatively little cost. Moreover, since he considered himself an "unknown quantity" as a scientist and wanted a chance to prove his abilities, Andrews offered to conduct the work without salary if the Museum granted him a leave of absence and allocated up to $1,000 from the Bowdoin Cetacean Fund to pay his expenses, terms to which Allen and Bumpus readily agreed.

Outfitted with notebooks, measuring tapes, binoculars, dissecting instruments, and a Graflex camera bought with his savings, Andrews left for British Columbia in May 1908. He spent the first two months at stations operated by the Pacific Whaling Company on Vancouver Island. Early in August, he moved to a facility owned by the Tyee Company on Admiralty Island, off the southern coast of Alaska. At both of these locations, several species of whales—humpbacks, blue whales, finbacks, and sperm whales—were hunted from fast, easily maneuverable steamers armed with harpoon guns. Using hoses connected to air pumps, the "kills" were inflated to prevent them from sinking, after which they were towed to the stations. When the carcasses had been hauled out of the water onto wharves, a crew of cutters or flensers stripped off the blubber with cables attached to winches and cut it into pieces that were boiled in cauldrons to extract oil used for lamps, candles, and cooking. Other men known as lemmers then sliced away the flesh, which was processed to make fertilizer, cattle feed, medicines, and glue.

Each time a whale was butchered, Andrews was able to take photographs and measurements of the animal, probe the contents of its stomach, and examine its internal organs and skeleton. He methodically dissected the uterus of every female, recording the size and weight of fetuses to determine the length of reproductive cycles. While aboard ship with the hunters, he kept meticulous notes on whales' behavior at sea, the length of time different species remained submerged, and the care given by females to calves. He even tasted whale's milk, pronouncing it "a bit too strong to be really pleasant." And because little was known about whales' reproductive habits, Andrews never missed a chance to observe their elaborate mating rituals, "spying on their love-making," he remarked, "like a Peeping Tom."

Over a four-month period, from June through September of 1908, Andrews studied a total of 106 whales. Added to this store of information were dozens of photographs and voluminous notes he had taken at sea. As an extra bonus, he purchased the skeleton of a humpback killed in Alaska, which he cleaned and shipped to the Museum. Altogether, he spent less than $700, an excellent investment of George Bowdoin's money, he felt, for one of the largest collections of data on Pacific whales ever assembled, most of it previously unrecorded.

Returning to the Museum in October, Andrews resumed a variety of duties in the mammalogy department. He also attended classes at Columbia University, where he enrolled to pursue a doctorate in zoology. The first course for which he signed up dealt with the evolution of mammals. It was taught by none other than Henry Fairfield Osborn, the lofty savant whom Andrews had secretly watched two years earlier from behind the meteorites in the Museum's entryway. Two or three evenings a week, Andrews also studied anatomy at the College of Physicians and Surgeons, which was later absorbed by the Columbia-Presbyterian Medical Center. As part of his course requirements in comparative anatomy at Columbia, Andrews was given access to the college's dissecting room, presided over by Dr. George S. Huntington, a prominent surgeon, teacher, and the editor for many years of the *Journal of Anatomy and Physiology*.

Under Huntington's tutelage, Andrews often worked in the laboratory until late at night. Like a scene from a horror film, he spent hours bent over a dissecting table illuminated by a single drop light, "surrounded by corpses in various stages of disrepair." Although he relished the chance to work alongside Huntington—who, impressed with Andrews' skill at dissection, encouraged him to consider surgery as a career—he found it impossible to adjust to the smell of human flesh, and to combat the odor, he became a heavy smoker. "A cigarette or pipe always had to be in my mouth," he wrote, "or I couldn't get through a long engagement with a corpse." Even so, he regarded his late-night encounters with cadavers as "enthralling," and came away from the experience with considerable medical knowledge that he would put to good use during his travels.

Apart from his busy schedule at the Museum and Columbia University, Andrews had also begun to pursue interests that would eventually lead to a lucrative career as a popularizer of natural science and exploration. Reports of his whaling adventures in the Pacific Northwest, illustrated with Andrews' photographs, had appeared in several newspapers around the country, including the *New York Times*. As a result, the widely read magazine *World's Work* asked him to contribute an article on shore whaling. Written in an informal, anecdotal style that would become a hallmark of Andrews' popular works, he was paid $250 for the piece—more than twice his monthly salary at the Museum.

The publicity surrounding his whaling exploits produced another bonus: it launched Andrews on his way to becoming a highly successful public speaker. At his first lecture, given before the New York Academy of Sciences in the fall of 1908, he delivered a lively commentary on a selection of spectacular lantern slides made from his photographs taken at whaling stations in Alaska and British Columbia. The evening was an unqualified triumph: Andrews received a standing ovation and an invitation to participate in a lecture program sponsored by the city's Department of Education.

After signing on, he learned to his dismay that the venue included not only schools and community organizations, as he had expected, but also jails and shelters for the homeless and alcoholics. His debut lecture, in fact, was held at the Five Points Mission in Lower Manhattan. Unaware that it was a refuge for derelicts, Andrews arrived in formal evening attire, complete with white tie and waistcoat, only to be greeted by a noisy, disheveled crowd, "men without coats or collars," he wrote, "and equally slovenly women." Even more upsetting was the presence of three policemen carrying nightsticks, who were stationed in the dingy hall to cope with riots that often erupted if the audience disliked the speaker. Afraid the inappropriateness of his clothing would generate hostility, Andrews quickly shed his coat, vest, and tie, complaining to the crowd that the room seemed overheated. He dispensed with his usual preliminary remarks and plunged into the slides, trying frantically "to make [the audience] feel and hear the rush of the sea, the roar of the harpoon gun, and the thrill of the hunt." Although he survived the evening unscathed, even drawing a smattering of applause, he later re-

called, "Never did I work harder, for there was the ever-present possibility of being plastered with overripe eggs."

Not long into Andrews' first season with the Department of Education, William Glass, who managed the prestigious J. B. Pond Lecture
Bureau, offered him a contract. The association with Pond quickly gave
him a wider lecture circuit, along with much-needed experience at a
professional level. Glass personally coached Andrews on ways to enliven his presentations and expand his repertory of subjects, and Andrews' talents on the podium soon catapulted him into the limelight as
one of the Pond Lecture Bureau's most popular speakers.

Andrews' eighteen-month apprenticeship at the American Museum had been remarkably successful, and there was ample justification for the high-spirited optimism expressed in a letter to his parents
on December 5, 1908, in which he raved: "I have never been happier,
worked harder, or felt more confident about the future." Andrews had
always harbored a compelling sense of predestination. Without a qualm,
he was now ready to set out in pursuit of whatever fate, or his "lucky
star," as he liked to call it, had in store for him.

CHAPTER 3

In June 1909, Andrews spent three weeks at Tadoussac in Quebec, near the junction of the Saguenay and Saint Lawrence rivers. He was sent there by the Museum and the New York Aquarium to collect specimens of the so-called white whales, a species of porpoise, highly prized for their snow-white fur, that migrated each spring from the Arctic region into the Saint Lawrence Seaway. After trying unsuccessfully to capture live examples for the aquarium, he acquired five specimens killed by trappers, enough material to provide the Museum with skeletons, plaster casts, skins, and photographs to use in preparing replicas of white whales for an exhibit of aquatic mammals.

Soon after Andrews returned, Bumpus summoned him to his office with a proposition. Would he be interested, the director asked, in going to the Dutch East Indies? Andrews' response was predictably exuberant. "I nearly leaped off my chair," he wrote in his autobiography. "It was ridiculous to ask me if I wanted to go anyplace. I wanted to go *everyplace*. I would have started on a day's notice for the North Pole or South, to the jungle or the desert. It made not the slightest difference to me."

Bumpus explained that a research ship, the *Albatross*, operated by the United States Bureau of Fisheries, was planning a cruise through the East Indies to study marine biology. Because of Andrews' experience as a cetologist, the Bureau had requested that he accompany the expedition to obtain information on South Pacific porpoises, a subject about which no scientific data existed. In addition, Andrews' duties would include going ashore whenever possible to gather mammals, birds, and reptiles for the Museum's collection.

Except for a few days spent purchasing special equipment, Andrews had no difficulty leaving on short notice. At the time, he was renting a room by the week at the Sigma Chi fraternity house at Columbia University, and he had already begun keeping an army trunk packed with essential clothing and ready for travel, a practice that started during his whaling junket to the Northwest and continued for many years. With the *Albatross* awaiting him in the Philippines, he booked passage on a Japanese steamer, the *Aki Maru*, which sailed for the Orient from Seattle.

From the moment Andrews stepped ashore in Yokohama, he was captivated by the spectacle that enveloped him: the teeming crowds, markets overflowing with exotic foods, ornate Shinto temples, carefully manicured gardens, "tiny women and children in their charming kimonos." Everything about the city's ambience overwhelmed him. "Even the smells, which most foreigners abhor, entice me," Andrews would later proclaim. "Almost before I knew it, I felt that I belonged completely to the Orient."

Accompanied by an English resident he met through mutual friends, Andrews spent his second evening in Yokohama visiting the celebrated Yoshiwara or "Street of Joy." Known throughout the Orient, the Yoshiwara was a stately, well-kept avenue lined on both sides with houses of prostitution. Each had a porchlike structure across its front enclosed by bamboo poles, which created the effect of cages. After dark, rows of girls, some no more than twelve or fourteen years old, dressed in kimonos and flooded by colored lights, arrayed themselves on white mats inside the cages where they could be seen by men strolling along the street. Every girl sat behind a small hibachi, which provided warmth and allowed her to prepare delicacies for passersby.

Outside the houses that catered mainly to foreigners, the prices were boldly announced on signs printed in English: SHORT TIME, THREE YEN; ALL NIGHT, FIVE YEN, INCLUDING BREAKFAST—about $1.50 and $2.50 at the prevailing rate of exchange. A prospective customer was free to converse with any girl he found attractive. If his interest progressed beyond that point, the financial arrangements were handled by the equivalent of a house steward, who then escorted the client to a comfortably furnished room where he was joined by the lady of his choice. Such was the Yoshiwara's fame that it became one of Yokohama's leading tourist attractions. Streams of visitors came to view its open display of "Oriental vice," and thousands of postcards were sold showing its girls, seated like brightly plumed birds in cages, waiting to entice customers.

Andrews' guide had arranged for them to dine at the most famous of the Yoshiwara's establishments, known simply as Number Nine. Unlike the other houses, Number Nine, which stood by itself at the end of the street, was two stories high and without the usual bamboo cages attached to its exterior. It boasted a lavish decor, a spacious courtyard, and a garden filled with enormous chrysanthemums. Its quiet, dimly lit rooms conveyed an aura of sybaritic pleasure; whatever one's desires, Number Nine accommodated them with incomparable style. Guests could linger over cocktails, enjoy music and dances performed by geishas, or feast on subtle dishes served at beautifully appointed tables. If a man cared to spend the night with a female companion, the price was 10 yen, or $5, with breakfast.

Apart from its felicitous atmosphere, superb food, and alluring women, one of the establishment's major attractions was its enigmatic owner, whom everyone knew only as Mother Jesus. Although she ran Number Nine with iron-handed efficiency, she had endeared herself to a wide circle of regular customers for whom she acted as adviser and confidante. Her clientele included spies, soldiers of fortune, smugglers, world travelers, ships' officers, and highly placed individuals in government, the military, and business. Mother Jesus was a clearinghouse of information who picked up gossip before anybody else, and could accomplish the seemingly impossible—such things as locating missing relatives and friends or arranging passage on ships supposedly booked to capacity.

Having heard about Mother Jesus ever since arriving in Yokohama, Andrews was nevertheless startled by his first impression of her. After dining on sukiyaki, served in a private room overlooking Number Nine's garden, Andrews' host, who had known the proprietress for many years, sent his calling card with a note asking her to join them. Minutes later, a screen slid back and Mother Jesus appeared in the flickering candlelight. "I don't know what I expected her to be like," Andrews recalled, "but certainly not what she was. She was barely thirty, slim and graceful, not beautiful, or even pretty, but strangely attractive with calm appraising eyes behind which seemed to lie the wisdom of the ages." In later years, Andrews' travels often brought him to Yokohama, and he and Mother Jesus became close friends. Each time he visited Number Nine, he took her gifts and they talked for hours about politics, mutual acquaintances, the latest intrigues, or his own adventures. "I not only liked but respected her," Andrews wrote. "But there was an aloofness and inscrutability about her . . . [and] one couldn't tell what she really thought or felt." Only once did he inquire into her past, but she revealed nothing except her age, the fact that she was born at Number Nine, and had never been married or taken a lover.

Reboarding the *Aki Maru*, Andrews proceeded to Shanghai, which he hated at first sight. Hong Kong, the *Aki Maru*'s next port of call, was altogether different, and Andrews proclaimed its azure-blue harbor, ringed by lush green hills, "one of the world's most spectacular sights."

After leaving Hong Kong aboard another vessel, the *Tamin*, he crossed the South China Sea to the Philippines, a three-day voyage marred by a typhoon and bouts of seasickness from which Andrews had suffered terribly ever since his whaling days in Alaska and British Columbia. He was scheduled to meet the *Albatross* at the Cavite Naval Yard near Manila, but when he arrived, the ship was cruising off Zamboanga and would not return for several weeks. Anxious to begin collecting zoological specimens, Andrews made arrangements through the secretary of the interior's office to visit an uninhabited island off the Philippine coast, reported to be a nesting ground for rare birds.

A week later, Andrews was lolling in a tropical paradise, accompanied by two Filipino boys he hired as assistants. The government steamer that delivered them had sailed away with instructions to pick them up in five days. Andrews delighted in the obvious analogy between his circumstances and the plot of *Robinson Crusoe*. Everything about the island had a storybook quality: wide sandy beaches fringed with palms, indigo-blue water broken near the shore by coral reefs, multicolored fish darting among the inlets, and lush vegetation echoing with the chatter of birds. Surveying the deserted island, Andrews conjured up the childhood fantasies that flooded his imagination each time his mother had read to him from Daniel Defoe's classic tale. Now it seemed as if those recollections had been transformed into reality, and he could almost envision himself as Robinson Crusoe. "Only instead of one, I had two 'men Fridays,'" he mused. What he did not yet realize was that, also like Defoe's protagonist, he and his companions would be castaways, for the ship failed to show up at the end of their five-day idyll, leaving them to survive by their own ingenuity.

If their abandonment posed any real danger, Andrews was oblivious to it. He set up a comfortable camp sheltered by large rocks, with hammocks suspended from trees, well out of reach of the enormous land crabs that roamed the beach at night. Nearby was a spring of freshwater, and aside from limitless supplies of fish and crabs, the island abounded with fat, black-and-white pigeons that were delicious when stewed or roasted over a driftwood fire. Every morning, Andrews was up at dawn, setting out after breakfast to explore the jungle, check his traps for small animals, and shoot new species of birds. In the afternoons, he and the two Filipinos swam in the surf, skinned and prepared the day's specimens, and collected marine life in tide pools.

Once his supply of shotgun shells ran out, Andrews—resourceful as always—instructed the boys to weave a net of palm fiber in which they caught pigeons and fish. With the addition of salt from evaporated seawater, they managed to eat a nourishing diet, despite the Filipinos' frantic concern that they would surely starve to death. When the steamer finally arrived after being delayed in port for two weeks with a broken propeller, its distraught captain, expecting to find the three beachcombers emaciated if not dead, discovered them feasting on pigeons stewed with palm roots and served in coconut shells.

On October 10, 1909, Andrews joined the *Albatross* at the Cavite Naval Yard. Lying at anchor, it impressed him as "a beautiful ship, with two masts, a single smokestack, and a wide afterdeck. . . . Outwardly at least it resembled a luxury yacht rather than what it actually was: the best-equipped vessel in the world for deep-sea exploration." Built in 1882 at a cost of almost $200,000, it was 234 feet long, weighed 384 tons, and its engines could achieve a maximum speed of just under ten knots. Although owned by the Bureau of Fisheries, the *Albatross* was manned by United States Navy personnel; it carried twelve officers, sixty crewmen, several biologists, and a Japanese artist who specialized in anatomical drawings of marine life. To facilitate its research activities, the ship was outfitted with two scientific laboratories, up-to-date equipment for oceanographic studies, and a variety of trawls, dredges, nets, and other devices that could be lowered beneath the water to retrieve specimens.

From Manila, the *Albatross* steamed southward through the Sulu Sea, stopping overnight at Tawau in British North Borneo (now Malaysia). Next, it proceeded to Sebatik, crossed the Celebes Sea, and began scientific operations off the northern tip of Celebes Island. Sailing leisurely along the coast, the ship anchored briefly at Menado before continuing south and west to Gorontalo, Ternate, Buru, Makassar, Ambon, Surabaja, and half a dozen other ports with equally alluring names, "places of pure enchantment," as Andrews called them—East Indian utopias, basking in a dreamlike ambience from which it seemed they would never awaken until they suddenly burst onto front pages as battlefields of World War II.

Except for periodic attacks of seasickness and his intense dislike of the hot, humid climate, Andrews savored everything about his voyage on the *Albatross*. He was fascinated by the enormous varieties of fish brought up from the ocean's depths, many with bizarre anatomical features designed to enable them to survive the complete darkness and crushing water pressure of their mysterious habitat. Above all, he enjoyed the time spent hunting and trapping on land each time the *Albatross* docked at ports to take on coal, or when the biologists were collecting off nearby islands. Throughout the cruise, Andrews saw only two or three schools of porpoises and no whales, so there was little for him to study at sea. But his zoological collection grew steadily until it included 50 mammals, 425 birds, and assorted reptiles and insects. Among his quarry were a new species of ant, a mouse, and an ibis, all of which were subsequently named in his honor—with the word *andrewsi* tacked on to their taxonomic designation.*

Yet the journey was not without its hazards. In north Borneo, trying to penetrate a dense wall of jungle, Andrews was severely gashed when he became entangled in ropelike vines bristling with thorns and had to

---

*In all, twelve previously unknown mammals, birds, reptiles, insects, and fish—including four extinct species from Mongolia—would eventually bear Andrews' name.

be cut free by machete-wielding natives. Once he suffered severe heat-stroke while trekking through heavy undergrowth when the temperature stood at 112 degrees in the shade. On the mountainous island of Buru, Andrews and two sailors stumbled upon hostile tribesmen who had abandoned their village minutes earlier, fled into the jungle, and left poisoned bamboo stakes concealed along the trail. In Borneo, a sharp-eyed guide alerted Andrews (who loathed snakes) to a twenty-foot python hidden in a tree, allowing him time to smash the snake's head with a bullet before he inadvertently walked under it. When the *Albatross* arrived at Formosa during the final weeks of the cruise, marauding bands of headhunters were terrorizing the countryside, preventing Andrews from going inland to collect animals. And a few days later, in the Formosa Channel, a typhoon struck the *Albatross* with such fury that her masts were cracked, scientific equipment was smashed, and the starboard engine was disabled, which required a week's layover in Keelung for repairs.

But these experiences did nothing to diminish Andrews' enthusiasm. On the contrary, his journals overflowed with the "interest and excitement" he felt every day of the cruise. "Each new port at which we docked, every tropical island, some rarely if ever visited by natives, the activities aboard ship, the vast expanses of ocean, all this kept me living in a dream world."

Andrews finished his work aboard the *Albatross* when it docked at Nagasaki in January 1910. He had originally planned to stay with the ship during its return voyage to the United States, but a casual visit to the city's market on his first day ashore precipitated a sudden decision to remain in Japan. Half a dozen shops were selling whale meat, and although Andrews knew that whales had long been a mainstay of Japan's food supply, he was unaware that shore whaling had been introduced recently to meet the demand. He realized at once that the existence of these whaling stations offered an extraordinary opportunity to expand upon the research he had initiated in the Pacific Northwest, particularly since whales from Japanese waters had never been studied systematically.

Armed with the Museum's permission and assurances of cooperation from the Japanese stations' owners, the Pacific Whaling Company, Andrews departed by steamer for a facility at Shimizu on the southeastern coast of Honshu Island. After waiting ten days without seeing a whale, he traveled from Shimizu eastward to a station in the village of Oshima, located at the entrance to Sagami Bay. Here he found an abundance of whales—finbacks, blue whales, humpbacks, killer whales, and seis—and it was not uncommon for two or three kills to be made every day. Because whales constituted such a high percentage of Japan's protein resources, in addition to providing oil, fertilizer, and baleen, the stations operated around the clock to prepare the meat for shipment to markets while it was fresh. The workmen carried out their gory task of butchering carcasses throughout the night, illuminated by oil lanterns that cast an eerie glow over the mist-shrouded wharves and reflected in pools of blood.

After a month at Oshima, Andrews journeyed north to the village of Aikawa on Sado Island in the Sea of Japan, an area visited by large herds of sperm whales, finbacks, blue whales, and seis. Once again, he labored tirelessly to secure measurements, anatomical descriptions, and photographs of as many specimens as possible, "often staying up all night," he wrote, "ghoulishly dissecting carcasses by torchlight. . . . Hardly a word existed in scientific literature regarding these whales that were coming under my observation in such numbers. It was a priceless opportunity." Andrews' research ultimately would net eighty tons of material, including ten porpoise skeletons and one each of a sperm whale, finback, blue whale, and sei. Along with skeletons, he boasted that his notes and photographs contained "more information on whales, living and dead, than had been gathered since whale hunting first began."

Apart from the considerable scientific rewards of his work in Japan, Andrews' travels around Japan in search of whales intensified his fascination with the Far East. His declaration, made a few days after arriving in Yokohama in September of 1909, that he already "belonged completely to the Orient," may have seemed impetuous at the time. But after living in Japan for nearly eight months, his exuberant first impression had evolved into a profound affinity with the country and a compelling urge to see more of Asia. Writing years later, Andrews acknowledged that his decision to stay in Japan in 1910 "probably changed the whole course of my life."

When he was not working at the whaling stations, Andrews immersed himself in Japanese culture. He ate nothing but native food, studied local customs, frequented marketplaces and craftsmen's shops, languished in teahouses, watched geishas perform, attended Kabuki theaters, and explored the countryside. Armed with a grammar and dictionary, he learned to speak passable Japanese. Eventually, he overcame certain traditions that first struck him as "exceedingly disconcerting": the exposed toilets shared by men and women alike, public baths that opened directly on to the street, and the almost complete lack of privacy and the unending curiosity to which he was subjected in the whaling villages. (At Shimizu, crowds punched holes through the rice paper walls of his hotel room and peered at him day and night.) Like most foreigners, Andrews was both amused and confounded by Japanese ideas of modesty—above all, their bathhouse etiquette, which dictated that "one might scrub a woman's back, or ask the same of her, but would never speak to her outside the bathhouse unless formally introduced." And he wrote of watching bare-breasted girls, standing in the doorway of a public bath, looking aghast at American magazines showing women in décolleté evening dresses, which they considered "frightfully immodest . . . and would never have worn."

Determined to sample every aspect of Japanese life, Andrews spent an evening in a foul-smelling opium den where his first experience with the drug left him violently ill. The effects were still worse when he was persuaded to try opium a second time by habitués who assured him that its continued use would induce delights "beyond the dreams of man." Yet as Andrews would soon discover, his reaction to other drugs was altogether different: he was "extremely susceptible to cocaine, heroin, morphine, and hashish," he confessed. "I tried them all to satisfy my insatiable curiosity . . . [and] learned to avoid them like the plague."

Andrews developed a genuine liking for the Japanese, and wrote of feeling "completely at home" once he adjusted to the inconvenience of being on public display. Everyone, even the ubiquitous onlookers, treated him courteously, "speaking, if at all, in whispers while maintaining their vigils." And the crews at the whaling stations were unfailingly tolerant of Andrews' research, though it often delayed their butchering procedures. As there were no doctors in any of the villages where he worked, Andrews treated a variety of disorders using the knowledge acquired during his studies at the College of Physicians and Surgeons. Eczema, venereal disease, and eye infections were the most common ailments, which Andrews remedied with liberal applications of zinc ointment and potassium permanganate. Twice he delivered babies, assisted by local midwives, and although he had no anesthesia, he pulled numerous teeth and amputated a man's hand that had been mangled in a winch.

While living in Aikawa, Andrews occupied a picturesque cottage overlooking the bay. Located in the station master's compound, it came with a "servant girl" named Kinu, who was, in Andrews' words, "tiny, pretty, delicate and eighteen years old." Besides cooking his meals, Kinu kept the house spotlessly clean, washed and ironed his clothes, tended the garden, and even nursed Andrews for two weeks when he was struck by a mysterious fever, never leaving him except to make pilgrimages to a temple to pray for his recovery. If Andrews failed to awaken at night when the station's whistle signaled the arrival of a whale, Kinu roused him, helped him into his oilskin coat and rubber boots, and led the way through the darkness to the wharf with a paper lantern. And each time he returned home after being away, even for short periods, Andrews found her kneeling at the door murmuring "polite phrases of greeting which, meaningless in themselves, nevertheless were pleasant to hear: 'I have been inexpressibly lonely while you were away. The sun has been under a cloud. The hours were dark. . . .'" Andrews regarded Kinu with great affection. He likened her to "a beautiful butterfly fluttering about in a flowered kimono . . . the huge bow of her *obi* (sash) always neatly arranged." He often surprised her with gifts, and when she confessed to owning only one flowered kimono, which Andrews preferred to her plain ones, he took her shopping and bought her three more. "From that time on she wore brilliant kimonos with reckless abandon, to the envy of all the village girls."

Despite the relative tranquillity of his sojourn in Japan, there were occasional brushes with unexpected dangers. At Aikawa, for example, a wounded finback capsized a boat in which Andrews and two Japanese sailors had rowed out from a whaling ship to deliver a *coup de grâce* with a lance. Just as one of the sailors plunged the steel blade into the whale's lungs, its tail suddenly arched upward out of the water and struck the boat, smashing one side and turning it over. Seconds later, Andrews and his companions found themselves swimming amid dozens of sharks, lured by blood gushing from the finback's blowhole, "a swarm of blue-black bodies [that] set upon the dying whale like a horde of vampires." Using pieces of wood from the splintered boat, the three men managed to fight them off until they were rescued.

At least that was one version of the mishap. There were others in which the details differed substantially. In a collection of adventure stories entitled *Heart of Asia*, published in 1951, Andrews tells of feeling "utterly terrified" as he kicked and flailed at the attacking sharks, one of which grabbed his foot and pulled off a rubber boot. Moments later, one of the Japanese sailors "shrieked in agony" as a shark tore away part of his calf muscle, leaving a wound that became so infected the leg had to be amputated below the knee at a hospital in Sendai. However, in his earlier book, *Ends of the Earth* (1929), Andrews treated the episode far less dramatically, stating that he "kicked one of [the sharks] in the nose but he only backed off and made no turn to bite." (The hapless sailor is not mentioned at all.) And in his 1943 autobiography, *Under a Lucky Star*, everyone escaped unscathed: "The sharks," he wrote apathetically, "weren't really interested in us."

Such discrepancies turn up elsewhere in Andrews' popular writings. Indeed, the shark episode—like several other harrowing experiences recounted in his nonscientific works—leaves no doubt that Andrews, who relished an exciting tale, occasionally used elements drawn from actual incidents to create semifictional stories tailored to a particular reading audience. It was hardly coincidental, for example, that the most hair-raising account of the shark encounter, in which the Japanese sailor lost his leg, originally appeared in 1949 in a publication intended for armchair adventurers, *True: The Man's Magazine*.

When the whaling season at Aikawa ended late in August, Andrews escorted the enormous crates containing his specimens to Yokohama, from where they were shipped to New York. Afterward, he lingered awhile in the city to rest and visit Mother Jesus at Number Nine in the Yoshiwara. He then traveled to Shimonoseki on the southern tip of Honshu, the headquarters for a fleet of fishing trawlers operating on both sides of the island in the Sea of Japan and the Inland Sea. Seeking an excuse to remain in Japan, Andrews had devised a plan to collect examples of interesting fish that showed up in the local market, an idea he sold to the Museum in a persuasive cable "asking for only three hundred dollars and guaranteeing fish by the score."

According to Andrews, Shimonoseki enjoyed the dubious reputation as "the hardest drinking port in the East." Any business that transpired, he observed, was "secondary to pleasure," and "wine, women, and song seemed to be the *raison d'être* among the foreign colony." Nearly every day the twenty or so Europeans and one American doctor, who lived in Shimonoseki, gathered at a club in the town's largest hotel and drank until "the servants 'poured' the gentlemen into their rickshaws in the middle of the afternoon and their offices saw no more of them that day."

Removing himself from this round of aimless socializing, Andrews settled into a secluded hillside house a mile north of town. The rent was 30 yen a month ($15) and included three servants, a well-kept garden, and a panoramic view of the harbor. In the afternoons, he watched the sun go down over the Inland Sea, slowly disappearing behind gnarled pines that enclosed the garden. He could observe the endless flow of shipping that crisscrossed the water: junks with bat-winged sails, square-rigged fishing boats, and trawlers. Now and then, ocean liners glided through the narrow channel into the bay, and in the evenings, Andrews often heard the ships' orchestras and saw passengers dancing on the afterdecks. At night, he fell asleep to the sounds of laughter, a plaintive bamboo flute, and the strumming of a samisen floating up from a teahouse at the bottom of the hill. "I was extraordinarily happy there," he recalled, wishing this Oriental idyll might never end.

Each morning, Andrews wandered through the markets alongside the docks to buy brightly colored fish from old women who tended the stalls. Soon the vendors came to know him and would put aside unusual specimens for his inspection. "Some even gambled," he wrote, "by buying strange fish that were not very good to eat in the hopes that they might suit my fancy." Almost every day, he acquired new species, including a variety of sharks, and within a few weeks he sent the Museum a sizable collection of marine life preserved in drums of formaldehyde.

Once this task was completed, Andrews no longer had any reason to stay in Japan. "Even my fertile brain," he admitted reluctantly, "could not conjur up another legitimate project." With most of his salary (now raised to $110 a month) having accumulated in the bank during the previous year, he decided to return home via Southeast Asia, Egypt, and Europe. From Shimonoseki, he went to Peking, the city that was to figure so prominently in his future. Here the Manchu Dynasty—now renamed the Qing Dynasty—still ruled over the last vestiges of its once-powerful kingdom under the boy emperor, Pu Yi, then five years of age, and his regent, Prince Chun. Beyond the sheltered confines of the court, an atmosphere of revolution was engulfing the country. By the end of 1911, the old regime was destined to collapse, undermined by the republican reformer Sun Yat-sen and his vision of a "regenerated China."

In the twilight of the Manchu's 268-year reign, Andrews caught a fleeting glimpse of its doomed splendors: stately processions that "seemed to step," he wrote, "from the pages of Marco Polo's journals"; lavishly attired nobles and their "gorgeous ladies with rouged cheeks"; the ornate temples, palaces, and courtyards of the Forbidden City, "that supreme testament," Andrews opined, "to China's glorious Imperial past." He marveled at Peking's bustling streets and alleys with their throngs of peasants, vendors, soldiers, foreigners of every nationality, rickshaw boys, and beggars. At the Hotel of the Six Nations, where he stayed for a week, Andrews mixed with its polyglot guests, "basking in an atmosphere of politics and diplomacy laced with thinly disguised intrigue." He delighted in rummaging through shops "overflowing with silks, embroideries, lacquer ware, brass, porcelains, ivory, jade, incense, and a bewildering array of foods, 'delicacies' such as snakes, bat wings, eels, insects, and dogs." Above all, however, Andrews was enthralled by the camel caravans arriving from China's western frontier, out of the vastness of central Asia, laden with goods from Tibet, Mongolia, Russia, Afghanistan, and India. "A hypnotic sight!" he exclaimed. "Undulating lines of camels, bellowing and snorting, driven by sunbaked Chinese and Mongols, their clothing and trappings caked with the Gobi's yellow dust."

Seduced by Peking's mystique, Andrews' infatuation with the Orient now reached a feverish pitch. Not content to be merely an interested traveler, he longed to become part of the compelling mosaic of life that surged around him. "Even then," he wrote, "with no idea of how or when, I vowed to someday live in this enigmatic city."

Next, he traveled to Shanghai and boarded a German ship bound for Singapore, where he checked into the Raffles Hotel and spent a week examining whale and porpoise skeletons in a local museum. Afterward, he proceeded to Kuala Lumpur and Pinang, "to gorge myself on mangoes . . . and the world's most delicious curry." He then booked passage across the Indian Ocean to Colombo in Ceylon. Frustrated to discover that Colombo "swarmed with rather dull Englishmen," he would remember it mainly for the moonstones "one bought by the cupful and gambled on finding a single stone worth the price of the lot"; and for the luxuriant garden of the Grand Oriental Hotel, with its palms bathed in colored lights, a fountain that splashed scented water into a stone basin, and an orchestra that serenaded guests as they lolled around tables placed beneath arbors covered with tropical flowers.

Sailing westward through the Red Sea and the Suez Canal, he landed at Port Said and caught a train to Cairo. Egypt's archaeological monuments so engrossed him that he engaged camels and a guide for a trip to El Faiyûm. Overwhelmed by even this brief encounter with the Sahara, Andrews noted in his journal: "Every vista was a revelation. The stupendous grandeur of the landscape, at once desolate, secretive, and forbidding yet supremely enticing . . . showed me that the desert was what I could love above everything else."

From Cairo, he journeyed to Italy "to sample the glories of classical ruins, Renaissance art, and Italian food." Afterward he walked over the Brenner Pass for a leisurely tour through Austria, Germany, France, and Belgium. Andrews' last stop was England, where he spent most of his time at Cambridge University and the British Museum of Natural History in London, conducting additional research on whales. When he finally reached New York in January 1911, aboard the *Kaiser Wilhelm der Grosse*, with exactly 5 cents in his pocket, plans for another trip to the Orient were already taking shape in his mind.

CHAPTER 4

During the seventeen months Andrews had been away from the Museum, a change occurred in its administration that would greatly affect his future. In 1910, his mentor, Hermon Bumpus, resigned as director. He was replaced by Frederick A. Lucas, a naturalist and authority on Atlantic whales who had spent twenty-two years at the Smithsonian Institution and seven years as curator-in-charge of the Brooklyn Institute of Arts and Sciences, now known as the Brooklyn Museum of Art. Throughout Lucas' thirteen-year tenure as director of the American Museum, he and Andrews maintained a cordial relationship, but it was now the Museum's president, Henry Fairfield Osborn, who assumed the key role in shaping Andrews' career.

Born in 1857 in Fairfield, Connecticut, Osborn had inherited a fortune from his father, William H. Osborn, a shipping magnate and president of the Illinois Central Railroad. As a student at Princeton, Osborn became interested in biology, geology, and paleontology. He spent the summers of 1877 and 1878 excavating fossil mammals in the Bridger Basin of southwestern Wyoming, an experience that caused him to abandon plans for a career in his father's business in favor of science.

After completing graduate courses in histology and anatomy at Bellevue Medical College in New York, Osborn went to England, where he studied embryology at Cambridge and enrolled at the University of London for a class in comparative anatomy under the celebrated naturalist and champion of Charles Darwin's theory of evolution, Thomas Henry Huxley. Returning to Princeton in 1880, Osborn completed his Sc.D. degree and was appointed assistant professor of natural science, teaching biology, comparative anatomy, and embryology. By the mid-1880s, he had become deeply immersed in vertebrate paleontology, especially the evolution of mammals, a subject that would emerge as his primary scientific interest.

In 1891, Osborn resigned from Princeton to accept a joint assignment to organize departments of biology at Columbia University and vertebrate paleontology at the American Museum. Osborn was asked to implement this formidable task because of his growing reputation as a scientist, his teaching and administrative skills, and, not least in importance, his family connections and wealth. An undertaking of this scope suited Osborn's protean ambitions perfectly. At Columbia, he inaugurated a wide-ranging curriculum in biology and zoology, together with a program of graduate studies in paleontology. As an integral part of his plan, Osborn combined the resources of Columbia and the American Museum for teaching purposes; in essence, the latter became an extension of Columbia by providing graduate students with access to the Museum's paleontological collections and allowing them to study under the supervision of its scientific staff.

At the same time, Osborn applied his prodigious talents to building the Museum's department of vertebrate paleontology. He hired some of the country's leading paleontologists, veterans and newcomers alike, such as Jacob Wortman, William Diller Matthew, O. A. Peterson, Walter Granger, and Barnum Brown. By dispatching teams to excavate rich fossil quarries in Kansas, Nebraska, Wyoming, Montana, western Canada, Egypt, and other localities, Osborn set about enlarging the Museum's collections of dinosaurs and extinct mammals. Aware of the public's intense fascination with long-vanished animals, especially dinosaurs, he initiated crowd-pleasing exhibits featuring mounted skeletons of such awesome creatures as *Tyrannosaurus*, *Brontosaurus* (now renamed *Apatosaurus*), and *Triceratops*.

When the long-reigning president of the Museum's board of trustees, Morris K. Jesup, died in 1908, he had already handpicked Osborn as his successor. Soon thereafter, Osborn relinquished his faculty position at Columbia (Andrews attended his final course) to devote himself exclusively to his responsibilities at the Museum—as chief administrator, fund-raiser, paleontologist, and, above all, as the architect of an unprecedented era of growth that catapulted the Museum to world-class eminence.

During Osborn's twenty-five-year tenure as the Museum's president, he greatly expanded the scientific and technical staff, extended the breadth of its research activities, and created new departments of herpetology, comparative anatomy, and ichthyology. Through his connections with New York's wealthy elite, he bolstered the endowment and enlisted influential friends and associates as trustees. Six buildings were added to the Museum's complex under his supervision, and a seventh—the Roosevelt Memorial Building, now the main entrance facing Central Park West—was begun while Osborn was president and opened a year after his death. Among his most ambitious achievements, and one that would pave the way for Andrews' rise to fame, was Osborn's role in setting in motion the Museum's "golden age" of exploration, a span of roughly a quarter-century that witnessed the launching of large-scale scientific expeditions to many of the earth's remotest corners, including what would become the crown jewel of this program—the Central Asiatic Expeditions.

A distinguished-looking man with a pronounced nose, piercing eyes, a shock of graying hair, and a neatly trimmed mustache, Osborn cut an imperious figure. He invariably dressed in sedate, three-piece suits with white piping along the edges of his vests, and he traveled about in a black limousine driven by a uniformed chauffeur. After establishing the Osborn Library on the Museum's fifth floor, he installed an imposing marble bust of himself gazing down upon the room, its features illuminated by a spotlight. His signature—written in a flourishing script with a broad-point pen—reflected his outsized personality; and when seated at his desk, he would majestically turn his entire massive body rather than just his head to greet a visitor. According to his own count, he belonged to 158 scientific, academic, and social organizations throughout the world, and was awarded twelve honorary doctorates from universities in the United States and Europe. Never one to minimize his accomplishments, Osborn compiled an autobibliography in 1934, listing 932 entries ranging from lectures, reviews, and speeches to articles and books. His masterwork, the two-volume *Proboscidea: A Monograph on the Discovery, Evolution, Migration, and Extinction of the Mastodonts and Elephants of the World*, required fifty-three years to complete. It was published posthumously, weighed forty pounds, and cost a staggering $280,000 to produce, a sum provided by J. P. Morgan, Sr., whose first wife happened to be the author's aunt.

Osborn's attitude toward the Museum was blatantly proprietary. He was notoriously single-minded in his determination to carry out his administrative policies, and his influence over the everyday running of the Museum often reduced the director's role to that of a yeoman. His peremptory tactics were a decisive factor in Hermon Bumpus' resignation in 1910, just as they would complicate Frederick Lucas' tenure in succeeding years. Osborn has been described as arrogant, tyrannical, insensitive, tactless, rude, and pompous. He also espoused controversial views on genetics, evolution, race, and other scientific issues that were frequently not shared by his colleagues.

In spite of his overbearing nature, Osborn often displayed flashes of humor, generosity, and engaging charm. And no one could deny that his impact upon the Museum was extraordinarily beneficial. The famed anthropologist Margaret Mead, who joined the American Museum in 1926, aptly described Osborn's influence in a quote cited by Geoffrey Hellman in a *New Yorker* profile: "[He] was a magnificent old devil," wrote Mead. "[The Museum] was his dream, and he built it. He was arbitrary and opinionated, but I got my first view of many things from his books and the exhibits he sponsored. Like Thomas Jefferson, he was a wealthy man who was also a scientific explorer. . . . We would never have had the Museum without him."

Over the years, Andrews and Osborn would develop a curiously symbiotic relationship, fostered largely by the fact that both men possessed an entrepreneurial spirit of sweeping dimensions. Osborn readily comprehended Andrews' flamboyant nature and believed in his ability to translate his ambitions into scientifically valid projects. Given his own fondness for gargantuan enterprises and his natural bent for showmanship, Osborn recognized early on that the aspiring zoologist-explorer's enthusiasm, energy, persuasiveness, and knack for public relations could benefit the Museum. Andrews, in turn, knew perfectly well that Osborn was the key to "making things happen," and he mastered the delicate art of stroking his new mentor's stupendous ego to his own advantage.

Such pragmatic considerations aside, it was clear that they genuinely liked and respected each other. At the outset, Andrews' admiration for Osborn, twenty-seven years his senior, bordered on hero worship, but in time their mutual bond gave rise to an almost filial attachment. Andrews joined the select few among Osborn's circle who were invited to the family's estate, Castle Rock, at Garrison in Upstate New York, where Osborn's father had built a fortresslike mansion overlooking the Hudson River opposite West Point. And during the years of the Central Asiatic Expeditions, when Andrews lived in Peking and only visited New York for brief periods, he sometimes stayed with the professor at his East Side town house, an honor seldom extended to anyone outside the family. Andrews once declared that Osborn was "truly a foster father"; and when Andrews' second son was born in 1924, he asked Osborn to be the boy's godfather, a request Osborn gladly accepted. The degree to which Osborn reciprocated Andrews' affection is revealed by a passage from a letter he sent to the then-famous explorer in China on December 8, 1924, in which he wrote, "I miss you greatly at my daily breakfast and walk across [Central Park]. You alone of all the men I know have a full measure of optimism; everyone else tells me things that cannot be done."

Early in 1911, soon after returning from his round-the-world sojourn, Andrews arranged a luncheon with Osborn. His purpose was to lay out an ambitious plan for a second trip to the Orient. While living in Japan the previous year, he told Osborn, he had learned of a whale called the *koku kujira* or "devil fish." Reportedly, they were hunted at only one location on the Korean coast, where they appeared every autumn and spring during their annual migrations to and from the Arctic. Andrews never actually saw a devil fish. Yet descriptions obtained from seamen who had killed these animals led him to suspect that they were a species believed by zoologists to be extinct—the California gray whale, which had once frequented the Pacific Coast of the United States and Mexico until hunters had supposedly exterminated them. Andrews was intent on visiting Korea to resolve the matter, assuming he could persuade Osborn to provide adequate funds.

But the search for devil fish was not his only objective. Andrews' interest in Korea had been further stimulated by the writings of Sir Francis Younghusband, the noted English explorer, soldier, and mystic. While on an expedition to Manchuria in 1879, Younghusband had climbed the jagged Paik-tu-san, or Long White Mountain (Paektu-san on modern maps), a nine-thousand-foot-high volcanic peak on the

Korea-Manchuria border that acquired its name from a thick layer of white pumice covering the summit. From its slopes, he had gazed across a primeval forest, cut by canyons and plateaus, that extended from the Paik-tu-san southward into Korea. Younghusband mentioned this forest briefly in his book *The Heart of a Continent* (1896), and, according to native informants, the region had never been explored. "No one could say whether it was inhabited," Younghusband stated, "or what secrets it concealed."

Intrigued by this uncharted wilderness, Andrews hoped to organize an expedition to penetrate its depths and collect zoological specimens. Before formulating his plans, he wrote to Younghusband, seeking advice on possible routes. "He replied that . . . if I could go in from the Korean side and make a traverse [of the forest] to the base of the mountain it would be a worthwhile job of exploration."

Osborn quickly saw the merits of investigating the devil fish and agreed to underwrite the project. But the Museum could only sanction the trek to the Long White Mountain, Osborn said, if Andrews raised the necessary money from private sources.

During the next eleven months, Andrews resumed classes at Columbia in pursuit of his doctorate in zoology. As time permitted, he lectured on his whale research and travels in Asia, and completed two papers for the Museum's *Bulletin* on porpoises he had collected in Japan. He also turned out an article on shore whaling for *National Geographic*, the first of three features he contributed to its pages over the years. Early in the spring of 1911, Andrews won another promotion, to assistant curator of mammals, but the pressure of new responsibilities at the Museum, added to lecturing, writing, and studying for his degree, often kept him working until past midnight.

Nevertheless, Andrews found time to secure the funding for his Korean journey by appealing to the ever-widening group of wealthy, well-connected people he was meeting through the Museum's trustees and benefactors. Everyone knew the venture had Osborn's approval, a fact that automatically gave Andrews powerful leverage. And by now he could capitalize on his own considerable assets as a fund-raiser. His whaling exploits had brought him widespread fame, and he had developed into a captivating public speaker with a gift for igniting enthusiasm for his travels. Witty and affable, with a commanding presence and inordinate good looks, Andrews, at age twenty-seven, was already being sought after in New York's social circles. Furthermore, his affiliation with the Museum allowed him to operate in what he called "an atmosphere of wealth and philanthropy."

In December 1911, Andrews departed from San Francisco on the maiden voyage of a Japanese ship, the *Shinyo Maru*. He reached Yokohama in time to attend a New Year's Eve party given by Mother Jesus at Number Nine—"one of her lavish extravaganzas, with masses of flowers, lanterns, and colored streamers festooning the house, tables piled with delicacies, and geishas dancing through the rooms accompanied by blindfolded flute players." Afterward, he traveled to a station owned by the Oriental Whaling Company at Ulsan on Korea's eastern coast, a "very merry place," he reported, "with . . . Korean sing-song girls flocking around the station." Because Andrews arrived at the peak of the hunting season, he did not have long to wait before encountering the mysterious devil fish. Just after sundown on his first night at the station, a shrill whistle signaled the approach of a kill. Dozens of workers lined the wharf in the bitter cold to watch the whaling ship, its deck shrouded with ice, glide quietly into the harbor. Suspended from its bow were the huge flukes of a devil fish.

Andrews observed that the flukes were shaped differently from any he had seen previously, and they bore peculiar markings in the form of circular gray blotches. As the creature was raised out of the water with cables, exposing the rest of its body, Andrews saw that the back was finless and there were longitudinal furrows along its underside. "Up came a wide, stubby flipper," he recalled, "then a short arched head. These told the story. It was the gray whale beyond a doubt." When two more specimens were killed the next day, he secured measurements and photographs, the first detailed pictures of these animals ever taken. During his six-week stay at Ulsan, he examined more than forty gray whales, studied their habits at sea, and acquired two skeletons.

Andrews also solved the quandary of the gray whale's apparent "extinction." Every year, large herds had once migrated from the Arctic Ocean southward along the Pacific Coast of the United States and Mexico in search of warm-water breeding grounds, particularly in the bays and inlets of California and the Baja Peninsula. By about 1880, gray whales had vanished from these haunts as a result of relentless hunting, and it was assumed the species had been killed off. Prior to Andrews' investigations, naturalists were unaware that a second migration route extended from the Arctic Ocean down the coast of Russia and Korea to the Yellow Sea, a region frequented by plentiful herds that had not been subjected to the intensive hunting practiced in American waters.*

Andrews' rediscovery of this allegedly extinct species was a scientific coup of considerable importance. Nevertheless, once the whaling season at Ulsan ended early in March, he could think of nothing but his journey to the Long White Mountain and he hastened to Seoul to prepare for the expedition. For generations, Korea, formerly known as the "Hermit Kingdom," had remained tightly closed to outsiders, ruled by feudal monarchs and a class of hereditary nobles called *yangbans* who wielded absolute power over the peasants. As a result of the Russo-Japanese War of 1904–1905, Korea had fallen under the domination of Japan, which annexed the country in 1910, set up a military government, and inaugurated a program of industrialization and trade. By 1912, the only unexplored part of the Korean peninsula was located along its border with Manchuria between the Tumen and Yalu rivers, the district in which the Long White Mountain was situated. In a 1919 article in *National Geographic,* Andrews described the region as one of "treacherous swamps, densely forested plateaus, and gloomy cañons— a vast wilderness treasuring in its depths the ghostly peak of the Long White Mountain, wonderfully beautiful in its robes of glistening pumice." Deep inside its crater lay a jade-green lake, the "Dragon Prince's Pool," discovered in 1709 by two Jesuit missionaries who reached the mountain from Manchuria and climbed to its northern rim.

Aided by the United States consul in Seoul and the Korean Bureau of Foreign Affairs, Andrews secured travel permits and letters of introduction to the commanders of military outposts along his route. He was also provided with an interpreter, a Japanese who spoke Korean and a little English—"a curious fellow," wrote Andrews, "who insisted on wearing an oversized frock coat and a badly ruffled silk hat throughout

*With the cessation of hunting in the 1880s, gray whales gradually reappeared on the Pacific Coast. By 1947, however, their survival was so endangered by renewed hunting that the United States placed them under federal protection.

the expedition." In addition, Andrews engaged a Korean cook named Kim, who at first tried to extort graft or "squeeze" by overcharging for food and pocketing half the money, until Andrews caught him in the act and delivered a stern reprimand, which included kicking him several times "like a football."

Early in April 1912, Andrews, Kim, and the interpreter left Seoul by train laden with camping gear, scientific equipment, and provisions. Boarding a ship at Pusan on Korea's eastern coast, they sailed north to Seshin (Ch'ongjin), then traveled by handcars along a railway to a village called Muryantei, where Andrews spent a "hellish" night in an inn swarming with "lice, bed-bugs, fleas, cockroaches, and spiders." Piling their baggage into three creaking oxcarts, the party proceeded to Musan, the largest city in northeastern Korea, located near the Tumen River just below the Manchurian border. Here the expedition lingered while Andrews attempted to track a man-eating tiger that reportedly had killed six children and numerous pigs, chickens, and dogs. Besieged, Andrews claimed, by terrified villagers who welcomed him "with open arms" and begged him to destroy the rampaging cat, he agreed to try. Assisted by a local hunter named Paik, who had earned the title of *sontair* for having single-handedly killed a tiger ("something like an English knighthood," Andrews noted, "for public service"), they spent three weeks searching the countryside within a radius of about fifty miles. But the elusive beast, which Andrews dubbed the "Great Invisible," always seemed to escape as if by some supernatural stealth.

Because of Paik's knowledge of the terrain, Andrews engaged him to remain with the expedition as a guide. At Musan, he had also expected to hire packhorses and drivers for his trek into the northern forests, but he found no volunteers for the journey. Everyone believed the area to be inhabited by demons, and it was rumored that bandits operated throughout the vicinity. Eventually, the problem was resolved with simple expediency by the commander of a Japanese garrison at Musan, a Lieutenant Kanada: he arbitrarily selected a group of men and ordered them to accompany Andrews under threat of harsh punishment. When Andrews' party finally set out from Musan, his entourage consisted of six heavily loaded horses, five *mafus* or drivers, Kim, Paik, and the Japanese interpreter.

After they passed through a section of the Tumen Valley, sparsely overgrown with oak and birch trees, the vegetation became more luxuriant and their progress was severely hampered by marshes, ferns, and moss-covered logs. With each mile, the forest grew thicker, and they soon encountered treacherous swamps made virtually impassable by a drizzling rain. At times, Andrews recalled, the "forest became so thick we had to cut our way through the tangled branches. . . . The Korean horsemen and in fact all the party came under the influence of the gloom and silence and it was difficult to force them to proceed."

As the altitude increased, the oak and birch trees gave way to larches sixty to one hundred feet in height, their branches covered with clumps of moss. The canopy of vegetation overhead was so dense that the woods were shrouded in an eerie half-light, pierced occasionally by shafts of sun. Although Andrews had expected to find the region teeming with wildlife, the forest was deathly silent. Unrelenting rain made walking a torturous ordeal, and in places trees had to be cut down, stripped of their branches, and laid end-to-end to provide footing for the packhorses. Sometimes, one of the animals would slip off these makeshift bridges and sink up to its stomach in mud, requiring frustrating delays while it was pried loose with poles. Moreover, the Korean drivers became increasingly disheartened, exhausted, and desperately afraid of losing their way. "The silence . . . of the forest began to work upon the imaginations of the Koreans," Andrews wrote, "and after we had been threading our way for five days through the mazes of an untouched wilderness the natives were discouraged and asked to return."

Just as the drivers' discontent threatened to erupt into a crisis, they caught sight of the majestic Paik-tu-san, the Long White Mountain, rising out of the forest. "Banked to the top with snow," Andrews observed, "it looked like a great white cloud that had settled to earth for a moment's rest. . . . The open sky and the mountain acted like magic on my men. They began to talk and sing and call to each other in laughing voices. . . . That night we camped in the shadow of the mountain . . . beside a pond of [melted snow]. I slept for fifteen hours, utterly exhausted."

Andrews was elated by the realization that he had at last penetrated the innermost recesses of the uncharted wilderness observed by Francis Younghusband more than a quarter of a century earlier. Because

Younghusband had already climbed the Paik-tu-san from the Manchurian side, Andrews was content to limit his explorations to the area around its base. Moreover, heavy spring snowdrifts at higher elevations would have made an ascent to the mountain's summit exceedingly difficult. Rather than attempting such an undertaking, Andrews spent a few days camping on its lower slopes.

Over vociferous objections of the Koreans, who wanted to return directly to Musan, Andrews decided to strike south and west to the Yalu River, thereby completing his traverse of the forest. Again their route carried the explorers through dense woods, swamps, and persistent rain. Occasionally, the procession stopped for much-needed rest or to collect zoological specimens. At one point, Andrews spent almost a week investigating three beautiful lakes, known to the Koreans as Samcheyong (now renamed Samjiyon), that glistened like jewels set amid a green larch forest, their shores outlined by gray volcanic ash.

At a village called Shinkarbarchin (Shin'galp'a-jiu), not far from modern Hyesan, Andrews dismissed his drivers and loaded his baggage aboard a log raft. Accompanied by his cook and the interpreter, he set out for Antung, at the mouth of the Yalu. After the five-month ordeal he had just completed, the 375-mile trip provided a restful interlude. As the raft drifted downstream, sometimes covering fifty or sixty miles a day if the current was swift, Andrews slept, fished, transcribed his notes, and shot ducks and geese, which he retrieved in a small boat tied behind the raft.

Throughout the entire expedition, Andrews had not seen another white traveler and had spoken only Japanese. Nor had he received any news of the outside world until he met an American missionary—a Dr. Gale—on the train to Seoul. Gale was aghast when Andrews told him of his plans to return from Asia via Europe and cross the Atlantic on a new ship, the *Titanic*. Andrews, of course, had no inkling of the April 14, 1912, disaster in which the liner sank on its maiden voyage.

By the time Andrews reached Lontag's Hotel in Seoul, dressed almost entirely in ragged Korean clothing and six weeks overdue, an enterprising journalist had circulated a report of the expedition's presumed demise in the wilds of the Manchurian frontier. Without waiting for confirmation, several newspapers had already published Andrews' obituary. He arrived to find the American Consulate and the Japanese

Foreign Office flooded with alarmed inquiries from relatives and colleagues regarding his safety. ("I have 'died' so frequently since," he later remarked, "that I am quite accustomed to it.") Andrews immediately sent cables to his parents in Beloit and to Henry Fairfield Osborn at the Museum, assuring them that he was alive and detailing the outcome of his journey.

Writing in *Under a Lucky Star*, he recalled that Osborn responded with a "joyous reply" in which he praised Andrews' accomplishments in glowing terms. "It was evident that the results of my first independent land expedition pleased [Osborn] enormously," he boasted. Yet in reality the venture had yielded relatively little of scientific value. Andrews himself declared the larch forest "a great disappointment from the standpoint of zoology." Except for a bear, roe deer, and wild boars, he saw no large mammals, and the traps he set out every night to catch smaller animals were almost always empty the next morning. Altogether, Andrews collected 162 mammals representing ten species, only two of which, a badger and a pika, were previously unknown to science; otherwise, his specimens consisted mainly of rabbits, hamsters, chipmunks, mice, and birds.

Despite its meager scientific bounty, the journey added substantially to Andrews' repertoire of hair-raising tales with which to enliven his popular writings, even if the line between fact and fiction was decidedly murky. One such incident involved a threatened mutiny by his Korean drivers when, deep in the forest, they became discouraged and afraid of losing their way. After Andrews refused to turn back, they hatched a plot, supposedly overheard by the interpreter, to flee during the night, taking the horses and food. Andrews valiantly tried to calm their fears, but nevertheless warned them that he and the interpreter would take turns guarding the camp at night. Anyone caught trying to leave was to be "shot without mercy." Yet published accounts of the incident are flawed by ambiguities, including whether or not the conspiracy was actually discovered before or after reaching the Paik-tu-san.

Then there is the matter of bandits. In his book *Ends of the Earth*, Andrews related how he stumbled into the camp of eight Manchurian brigands—"tall, brown, hard-bitten fellows armed with flintlock rifles." Relying on a few words of Chinese he had learned, he allegedly talked his way out of the potentially dangerous situation by inviting the bandits to inspect his camp (to prove that he was transporting nothing of value) and asking them to stay for dinner. But there are puzzling inconsistencies regarding how much of a threat bandits really posed. In a letter to a friend, George Borup, written on March 14, 1912, just before the start of his expedition, Andrews stated that Japanese officials were "making a bit of a row about my going alone," since the area supposedly swarmed with bandits. Writing in the *American Museum Journal* soon after his return from Korea, Andrews inexplicably dismissed the threat of brigands, denouncing as "absurd" reports that robbers operated in the region along the border. Seven years later, however, in his 1919 article for *National Geographic*, he expressed a sober view of the subject by claiming that "wandering gangs of Chinese and Korean bandits have ranged along the forest borders, keeping the natives in terror."

The encounter with bandits, the foiled mutiny in the Korean wilderness, and the tiger hunt at Musan, which later turned into a highly embellished quest for the "Great Invisible," all ended up in Andrews' popular books and articles as partly romanticized, but to what extent it is difficult to determine. Undoubtedly, though, as with the shark incident in Japan two years before, they are clear-cut examples of his liberal use of literary license as a means of appealing to general readers; moreover, such suspenseful stories greatly enhanced his adventurous image.

After departing Korea, Andrews revisited Peking, lured by memories of his first visit to the city. Only this time he found China seething with unrest. By the end of 1911, the Manchu Dynasty had been toppled from power. Its last emperor, Pu Yi, was confined to the Forbidden City, a pathetic victim of events over which he had no control. Outside the imperial refuge, itself ridden with intrigue, the fledgling republic created by the recent revolution was hovering near collapse under the leadership of the Machiavellian Yuan Shih-kai, a self-serving army commander to whom China's idealistic leader, Sun Yat-sen, had unwisely yielded the presidency of the new regime.

Unperturbed by the gathering political storm, visitors to Peking continued to frolic among the pleasurable diversions enjoyed by members of the foreign colony. Andrews was no exception. He dined at the city's best restaurants. At the Hotel of the Six Nations, he sipped cocktails, learned to play mah-jongg, and danced to Western-style orchestras. He attended legation parties, shopped for souvenirs, explored famous landmarks, and visited the Great Wall. Like most foreigners, he learned to steel himself against the civil unrest and atrocities that occurred almost daily. Even the public executions aroused a morbid fascination as he watched the open carts that continually passed through Peking's streets carrying prisoners, stripped to the waist and tightly bound on their way to be executed in a square near the Temple of Heaven. "It was not a pleasant sight, those head-lopping parties," Andrews wrote, "but the people seemed to love them judging by the crowds that followed the tumbrils on their gruesome journey." Joining a throng of bystanders at one execution, he reported that eighteen men lost their heads in a few minutes' time. "They kneeled in a long line and the executioner with his broad-ended sword started work as nonchalantly as though he were chopping trees."

Leaving Peking after nearly a month, Andrews went to Russia via the Trans-Siberian railway at the invitation of a prince he had met in Japan in 1910. For over six weeks, he remained at his host's lavish estate outside Moscow, languishing in "sinful comfort." He remarked, ". . . One cannot imagine such luxury, such total disregard for money, such pampering of every whim as characterized the nobility before the revolution. . . . We rode and shot, danced and played, drank and ate— particularly did we eat. Foods such as I have never dreamed existed, with vodka and caviar served in huge blocks of ice, seemed always on the table. . . . Of course I fell violently in love (or thought I did) with a beautiful Russian girl who reciprocated my affection as a matter of hospitality and promptly forgot me when I left."* Eventually, Andrews confessed, his conscience became "bothersome" and he reluctantly abandoned his dalliance in Russia and started home, although not be-

---

*Apparently, Andrews never identified his Russian host in any of his writings, although among his private correspondence there is a letter dated December 17, 1933, from a Georgian prince, Alexis Mdivani, whom Andrews appeared to have known for years. As for the object of his affections, he made only veiled references to a young woman known as Olga. And according to information obtained by the author from Andrews' second wife, he often spoke of having been received while in Russia by the Grand Empress Marie, the mother of Czar Nicholas II.

fore spending another two months wandering through Finland, Sweden, Norway, and Denmark.

It so happened that Sir Francis Younghusband visited the American Museum while on a lecture tour of the eastern United States in the spring of 1914. At a reception in his honor, Andrews had an opportunity to tell Sir Francis about his expedition to the Long White Mountain and the forbidding forest that blanketed its southern approaches. Younghusband was enthralled by Andrews' firsthand account, and their conversation marked the beginning of a friendship that lasted until Younghusband's death in 1942. "It had been his book and encouragement," Andrews wrote with his ever-present belief in destiny, ". . . that had sent me on my first land expedition and forever changed the direction of my life."

PART TWO
The Central Asiatic Expeditions: The Inspiration

Always there has been an adventure just around
the corner—and the world is still full of corners!

—Roy Chapman Andrews

CHAPTER 5

Andrews returned from Korea to find himself a celebrity in the making. At age twenty-eight, he was a recognized authority on Pacific cetaceans, and his widely publicized Korean journey had established his credentials as an explorer. When he arrived in New York, he was inundated with requests for interviews, lectures, and articles, some of which promised lucrative financial rewards. Andrews wisely refused to allow these inducements to sidetrack him from completing his graduate studies at Columbia. Until now, he had pursued them on a "hit or miss basis; mostly miss," he wrote, "for I hadn't been able to stay put long enough to do consecutive resident work." Aware that his schedule no longer provided sufficient time to earn a doctorate, as he had originally intended, he decided to complete a master's degree instead. Using his material on gray whales as the subject of his thesis, he worked feverishly for six months, turning the research and photographs he had collected at Ulsan into a scholarly treatise that he submitted in time to receive his master of science degree in June 1913. Entitled "The California Gray Whale (*Rhachianectes glaucus* Cope): Its History, External Anatomy, Osteology and Relationship," it was published in 1914 as part of the Museum's *Memoirs*.

Midway through the summer of 1913, Andrews was off on another trip aboard a luxurious yacht called the *Adventuress*. Joining the boat's owner, John Borden, and a group of his friends, Andrews set out from Long Island bound for Alaska and the Chukchi Sea via Cape Horn. Borden, a wealthy Chicago businessman, lawyer, and sportsman, had built the *Adventuress* at a cost of $50,000, outfitted it with special equipment, and conceived the idea of harpooning a bowhead whale as a gift to the American Museum. Since the bowhead, an Arctic-dwelling species, was the only whale not represented in the Museum's collection, Andrews went along on the voyage to supervise the preparation of the specimen Borden expected to kill.

From the beginning, the venture was a disaster. Because the *Adventuress* was delayed in rounding Cape Horn, it ran into impenetrable ice flows near the Arctic Circle. To further complicate matters, the friends Borden had invited along on the cruise, a disagreeable, hard-drinking bunch, had expected the excursion to be a continuous party rather than a serious scientific endeavor. Andrews, frustrated by the situation, described the passengers as "less suited for the expedition than the yacht," concluding disgustedly that "boating parties and science do not mix."

Although Borden failed to kill or even photograph a bowhead whale, Andrews was able to salvage something of zoological value out of the fiasco. He collected waterbirds, caribou, and mountain goats along the coast of Alaska; and while hunting on Kodiak Island, he shot three Alaskan brown bears. On the return voyage, Borden dropped Andrews off at Saint Paul, one of the Pribilof Islands in the Bering Sea. At the request of the Bureau of Fisheries, he spent three weeks on the desolate, wind-swept beaches studying the huge seal herds that converged there during the mating season, when a single rookery often teemed with thousands of seals. By the time he left Saint Paul on a Coast Guard cutter bound for Alaska, where he rejoined the *Adventuress*, his research had produced a substantial collection of anatomical data, still photographs, and some of the most detailed footage of seals ever captured on film.

Back in New York again, Andrews plunged into curatorial duties in the mammalogy department. Among various projects, he completed a monograph on sei whales that was printed in the Museum's *Memoirs* in 1916. In addition, he contributed papers to the Museum's *Bulletin* and *Journal*, and turned out several magazine articles. At the urging of an editor with the publishing house of D. Appleton and Company, he also set to work writing a book about his whaling experiences, *Whale Hunting with Gun and Camera*, which appeared in 1916. A lively blend of scientific facts, firsthand observations, and raw adventure, it was served up in his usual anecdotal style that garnered respectable reviews, though its sales were disappointing.

With no immediate plans for another expedition, this was the first time since 1908 that Andrews remained at the Museum long enough to engage in the traditional duties of a curator on anything like a sustained basis. But tormented by his innate restlessness, the often tedious minutiae of curatorial responsibilities eventually proved impossible to reconcile with his temperament. "Not that I didn't like Museum work," he asserted. "On the contrary, I loved it. The Museum had come to be a part of [me], and there was no phase of [its] activities . . . that did not fascinate me. But I loved wandering more. . . . The lure of new lands, the thrill of the unknown, the desire to know what lay over the next hill!"

Nor, it must be remembered, was Andrews an intellectual in the true sense—something he freely admitted. He was attracted to the *idea* of science and the opening of hitherto unknown scientific horizons, but his interest in the deeper implications of his discoveries was often superficial. As Douglas Preston observed in his history of the American Museum, *Dinosaurs in the Attic*, most great explorers were "indifferent scientists." Andrews was no exception. He thrived on blazing pathways into unknown places, but he was usually content to leave the interpretation of whatever scientific bounty his journeys yielded to specialists equipped with the training and patience he lacked.

Given these considerations, it was not surprising that Andrews' Korean expedition in 1912 marked a significant turning point in his career. "Ever since [then]," he acknowledged in *Ends of the Earth*, ". . . I had been sure that [land exploration] was what I could do best and what would make me happiest." By 1916, Andrews had decided to abandon his pursuit of cetaceans. He was now engrossed in studying Asian land mammals, but like his involvement with cetaceans, this specialty, about which he acquired a vast amount of knowledge, would never become his life's work either. He had long since realized that zoology was secondary to his deeper obsession: it supplied him with a passport to adventure. As his whaling research and Korean explorations had demonstrated, zoology afforded legitimate reasons to venture into exotic corners of the earth that had always attracted him.

Zoology aside, in *Under a Lucky Star*, Andrews offered a more candid explanation for his restive nature:

During all my years at the Museum I almost never returned from an expedition without having plans ready for another. I had found it wise to strike while the iron was hot. The enthusiasm of returning from a successful trip carried great weight in the plans for a new one. Then again, I was afraid I would get so immersed in Museum affairs that the authorities might think I had better stay home for a while. If the new trip had been approved, and was in the offing, it was much easier to keep my particular decks cleared for action. . . . Sometimes when I walked across [Central Park] on a starlit summer night I used to look up at the drifting clouds, going with them in imagination far out to sea into strange new worlds. Then I would count the days that still remained before I could set my feet upon the unknown trails that led westward to the Orient. . . .

As Andrews' fame magnified over the years, such sentiments would alienate a small but vociferous group among the Museum's staff, mainly practitioners of the low-profile, tedious but vital scientific research that Andrews found too restricting. These detractors tended to characterize him as opportunistic, headline-grabbing, and a self-aggrandizing adventurer. Yet whatever reservations his critics expressed regarding their flamboyant colleague, Andrews' pursuit of "unknown trails" was no mere romantic metaphor: it literally placed him in a position described by the New Yorker's writer Geoffrey Hellman as "the Museum's most celebrated . . . explorer."

Being a full-time resident of New York since his return from Korea in 1912 greatly benefited Andrews' social life. When he was not attending receptions, banquets, and lectures at the Museum, he frequently went to the theater and concerts, or was invited to dinner by some of the city's most prominent hostesses anxious to cultivate the much-talked-about explorer. On weekends, Andrews often visited friends in Connecticut, Upstate New York, or Long Island, where he hunted, rode horses, and socialized with influential families. During this time, Andrews also learned to play polo, a pursuit to which he became passionately addicted. Exhibiting a natural flair for this fast-paced sport, Andrews mastered its intricacies with formidable skill, and while living in Peking years later, he would earn accolades as a topflight player.

By now, the debonair young explorer had also acquired a much-deserved reputation as a "ladies man." His eye for the opposite sex was well cultivated, and he was frequently seen in the company of attractive women he met in New York's social circles or during his sporting weekends in the country. Yet there seems to have been no time in his peripatetic schedule for anything other than fleeting romantic involvements with nameless and, one suspects, quickly forgotten lovers.

Everyone was stunned, therefore, when in the fall of 1914 Andrews announced his engagement to Yvette Borup, the beautiful, high-spirited, and talented sister of his late friend George Borup, who had drowned in a boating accident while Andrews was in Korea. Yvette and George were the children of Major Henry Dana Borup of Saint Paul, Minnesota, and Mary Watson Brandreth, a native of Ossining, New York. After graduating from West Point and serving as an ordnance officer in the Indian Territories of Nevada and Texas, Major Borup was assigned to a variety of legation posts in Europe, including London, Paris, Berlin, and Potsdam.

In 1891, Mary Borup gave birth to Yvette in Paris; four years later, their son, George, was born in Ossining while his mother was visiting relatives. After attending Groton, George graduated from Yale in 1907 with a degree in geology. While working as an assistant in the American Museum's department of geology—where he became fast friends with Andrews, newly arrived from Beloit in 1906—Borup met the famed Arctic explorer Robert E. Peary. Impressed with Borup's intelligence, knowledge of geology, and superb physical condition, Peary, who was preparing for his historic polar expedition of 1908–1909, selected Borup to supervise the supplying of forward base camps used by Peary to reach the North Pole on April 6, 1909, an adventure Borup recounted in his book, *A Tenderfoot with Peary.*

Tragically, on April 28, 1912, while planning another polar expedition with Donald B. MacMillan, Borup and a fellow Yale alumnus, Samuel Winship Case, overturned and drowned while canoeing in Long Island Sound. An eyewitness reported that Borup, an accomplished swimmer, apparently died in a valiant attempt to rescue Case, but the details of the mishap were never explained.

Although Yvette possessed a streak of her brother's adventurousness, she spent a rather frivolous youth as a debutante in the highest echelons of European society. Molded by her Parisian upbringing and extensive travels on the Continent, she radiated an engaging charm. Slender and graceful, with chestnut hair and flashing brown eyes, she was fluent in French, German, and English. Early on she studied ballet and drama at the Paris Conservatory. When her father was transferred to Germany, she was enrolled in the Kaiserin Auguste Stift in Potsdam, founded by the Hohenzollerns as an elite school for the children of aristocratic parents.

Here Yvette mingled with Germany's haut monde, including Princess Victoria Louise, Kaiser Wilhelm's only daughter. Soon the two girls were almost inseparable, and Yvette became a frequent guest at the German court. So close, in fact, was her friendship with the princess that she joined a select contingent of Americans invited to attend Victoria Louise's marriage on May 24, 1913, to Prince Ernest Augustus of Brunswick, one of Germany's richest noblemen and the Duke of Cumberland's heir. A dazzling affair, the wedding, with its complement of military pomp and grand balls, was among the last of the Continent's royal spectacles—attended by King George V and Queen Mary of England, Czar Nicholas II of Russia (who arrived aboard his lavishly appointed train under heavy police protection), and a bewildering array of nobility from all over Europe. Worsening political tensions would force Yvette to return to the United States after the wedding, but despite the outbreak of World War I, she and Victoria Louise remained lifelong friends.

How Andrews and Yvette met is uncertain. Some sources have erroneously stated that they first encountered each other at her brother's funeral, but this was impossible, since Andrews did not return from Korea until months after George Borup's death. The most likely explanation—one that appeared in various newspaper stories at the time of their wedding—is that Yvette, recently back from Europe in 1913, was introduced to Andrews by mutual friends to whose home they were invited on the same evening. Regardless of the circumstances that brought them together, within a matter of days their courtship was flourishing.

On October 7, 1914, they were married at Trinity Episcopal Church in Ossining, New York, amid a flurry of press notices and Sunday-supplement features touting an idyllic love match between the petite, glamorous, twenty-three-year-old socialite and the intrepid explorer whose deeds had ignited the public's admiration. After the wedding, the couple settled in a comfortable house in Lawrence Park, Bronxville. It was the first quasi-permanent residence Andrews had known since leaving Beloit, having lived in rented rooms, apartments, hotels, and the Sigma Chi fraternity house at Columbia. But this newfound domesticity was destined to be short-lived. Within eighteen months, Andrews and Yvette would embark on a grueling expedition to southwestern China, a region convulsed by revolution and overrun with bandits.

Seeking a reason to initiate further explorations in Asia, Andrews turned his attention to a concept set forth some years earlier by Henry Fairfield Osborn. For almost two decades, Osborn had been studying the evolution and distribution of mammals, both living species and their extinct precursors. His research led him to formulate an hypothesis that fired Andrews' imagination and supplied the impetus for launching an innovative new enterprise.

On April 13, 1900, the journal *Science* had published an article by Osborn in which he expounded his theory. Based upon the comparative anatomy and geographical distribution of various species of mammals, Osborn concluded that Asia had been the origin point for most present-day mammals and the place from which they had migrated to Europe and America. To support this idea, he cited the fact that animals such as the elk, moose, bighorn sheep, caribou, and mountain goat found in Europe and the western United States were closely related, suggesting that their genesis had occurred somewhere between these areas—namely, in Asia. In the foreword to Andrews' 1926 book, *On the Trail of Ancient Man*, Osborn paraphrased the concept he had outlined in his article for *Science* as follows:

> First, on opposite sides of the globe we observe two great colonies, one in Europe and one in the Rocky Mountain region of America, which are full of different degrees of kindred in their mammalian life; yet they are separated by ten thousand miles of intervening land in which not a single similar form is found.
> The fact that the same kinds of mammals . . . appear simultaneously in Europe and the Rocky Mountain region has long

been considered strong evidence for the hypothesis that "the dispersal centre is half-way between." In this dispersal centre, during the close of the Age of Reptiles and the beginning of the Age of Mammals, there evolved the most remote ancestors of all the higher kinds of mammalian life which exist today. . . . The history of northern Asia remains unknown until the period of the Ice Age, when man first appears; yet theoretically we are certain that it was part of a broad migration belt which at one time linked together the colonies of France and Great Britain with those of . . . Wyoming and Colorado. Though the kinds of animals which we find in these two far-distant colonies are essentially similar . . . connecting links are entirely unknown. It follows that northern Asia must [have been] the migration route between these two colonies.

Equally important, Osborn identified northeastern Asia as the locale where the forerunners of humans—the primates—had evolved, a belief that prompted him to designate this region as the "Cradle of Mankind." Notwithstanding the lack of empirical evidence to verify his assumption, Osborn was certain that proof would be discovered eventually, and his reputation was such that his theories could not be dismissed summarily.

Several years later, another towering figure in the field of mammalian evolution, William Diller Matthew, embraced this thesis. Matthew had joined the Museum in 1895 as one of the bright lights recruited by Osborn for the department of paleontology. A brilliant evolutionary theorist and prolific author, Matthew published a classic work, *Climate and Evolution* (1915), in which he argued that mammals had migrated out of Asia to far-distant parts of the globe by means of land bridges such as the Bering Strait between Asia and North America, the Isthmus of Panama into South America, and now-vanished connections leading from Europe to Africa and Southeast Asia to Australia.

Like Osborn, Matthew was convinced that the dispersal center for primates and their human offshoots lay somewhere on the Asian continent, "probably in or about the great plateau of Central Asia." Furthermore, he envisioned this area not only as the home of mankind's primordial ancestors but also as the birthplace of civilization itself:

In [the central Asian plateau], now barren and sparsely inhabited, are the remains of civilizations more ancient than any of which we have record. Immediately around its borders lie the regions of the earliest recorded civilizations—of Chaldea, Asia Minor, and Egypt to the westward, of India to the south, and of China to the east. From this region came the successive invasions which overflowed Europe in prehistoric, classical, and medieval times, each tribe pressing on the borders of those behind it, and in turn being pressed on from behind. The whole history of India is similar—of successive invasions pouring down from the north. In the Chinese empire, the invasions came from the west. In North America, the course of migration was from Alaska, fan-wise to the south and southeast and continuing down along the flanks of the Cordilleras to the farthest extremity of South America.

Andrews had first read Osborn's article in *Science* while studying at Columbia in 1909, and he was undoubtedly exposed to a thorough examination of its premise during Osborn's lectures. With the appearance in 1915 of Matthew's *Climate and Evolution*, a far-reaching plan began to take shape in Andrews' mind. He resolved to test the validity of the Osborn-Matthew theory by searching for scientific evidence that Asia had been the "Garden of Eden" for mammalian life, including the ancestors of man. The idea was tailor-made for a zoologist-turned-explorer who yearned for a new challenge.

Once his concept was formulated, Andrews, supremely confident of its merit, approached Osborn with his most sweeping proposal to date: a series of expeditions, extending over several years, aimed at collecting Asian animals and determining, if possible, the relationship to their counterparts elsewhere in the world. Eventually, an attempt also would be made to uncover the fossil record from which the evolutionary history of these animals might be reconstructed. The project's initial phase involved a collecting trip to the southern edge of the central Asian plateau—namely, Yunnan Province in southwestern China and along the Tibetan frontier. Its purpose was to allow Andrews to acquire animals from an area virtually unknown zoologically, and to familiarize himself with the country in preparation for subsequent trips into central Asia itself.

Osborn could hardly have failed to give his blessing to the plan. Andrews was proposing a grand-scale effort to supply the key element that his theory of mammalian evolution lacked—actual scientific verification. It was agreed that the Museum would provide half the $15,000 needed for the expedition if Andrews raised the balance. With the weight of Osborn's prestige and social contacts firmly behind the project, the deficit was made up in a few months' time. All the contributors had previous ties to the Museum, either as trustees or patrons: George S. Bowdoin, whose Cetacean Fund had financed Andrews' whale research; the steel magnate Henry Clay Frick and his son Childs Frick, both prominent benefactors of the Museum; James B. Ford; Mr. and Mrs. Sidney M. Colgate, whose family founded the soap and perfumery empire; Mrs. Adrian Hoffman Joline; the textile manufacturer Charles L. Bernheimer and his wife; and Lincoln Ellsworth, the celebrated polar explorer, heir to a coal-mining fortune, and a close friend of Andrews' late brother-in-law, George Borup.

By the beginning of 1916, preparations were complete for what the Museum officially designated as the Asiatic Zoological Expedition. Andrews, of course, was the leader and zoologist. Arrangements were made for Edmund Heller, a naturalist with experience in Africa, South America, Mexico, and Alaska, to join the expedition in China to collect and prepare small mammals. Yvette went along as the photographer—a logical choice, interestingly enough, since she had been deeply engrossed in photography for years, learning the basics from her father before studying with teachers in France, Germany, Italy, and New York. She had mastered black-and-white prints, relatively new color techniques (especially the use of Paget color plates, which produced remarkably realistic results), and motion pictures. Until then, Yvette's photographic work had been the avocation of a talented amateur, and she had certainly never experienced anything like the physical hardships of exploration. Equipped with two Kodaks, a tripod-mounted four-by-five Graphic, a Universal movie camera, and a Graflex, plus a large supply of film and plates sealed in airtight tins and a collapsible rubber darkroom, she plunged into the rigorous, nineteen-month journey that awaited her without the slightest trepidation.

By the time the expedition was ready to depart, a pall of apprehension was cast over the trip by the outbreak of war in Europe and rapidly deteriorating political conditions in China. Yuan Shi-kai, the former military strongman, had assumed the presidency of the infant republic on February 15, 1912. But Yuan harbored grander ambitions. Never a true republican at heart, he had covertly plotted to resurrect the monarchy and install himself as emperor. Having orchestrated the nullification of the republic, Yuan ascended the Dragon Throne on December 12, 1915, although these high-handed tactics proved widely unpopular and revolution broke out almost immediately. On December 25, Yunnan Province—Andrews' destination—declared its independence from Peking, followed shortly by the neighboring provinces of Kweichow and Kwangsi. Soon the entire country appeared to be hovering on the brink of rebellion. Worried about the safety of the expedition, the Chinese minister to the United States, Wellington Koo, met with the Museum's representatives to discuss the situation. Koo strongly urged postponing the venture, but Andrews was determined to proceed regardless of the risks.

On March 28, 1916, he and Yvette sailed from San Francisco aboard the SS *Tenyo Maru* bound for Japan. They left amid an even greater outpouring of publicity than had surrounded their marriage. "We went off to the accompaniment of popping flashlight bulbs and screaming headlines," Andrews recalled. Reporters besieged them at every stop across the country, crowding onto the train to bombard them with questions. One writer demanded to know what the fashion-conscious Yvette planned to wear in the bush, and whether she would carry a weapon. A perturbed Andrews replied that she "expected to wear exactly what the men wore"; that she would "indeed carry a rifle and not hesitate to use it." Even in China, they abandoned all hope of escaping the onslaught of sensation-hungry reporters; and several English-language newspapers, including the *China Star* in Shanghai, took a cue from American papers and ran articles about the famous couple's perilous "honeymoon," illustrated with absurdly romanticized drawings of Yvette as a demure bride in her wedding dress and veil.

As if the journalistic frenzy over the couple's daring trek into the bandit-infested wilds of Yunnan was not enough, there was another sidelight to the story with which to lure readers: the "blue tiger," an elusive man-eater, strangely colored with blue-gray fur and black stripes, that was terrifying peasants in Fukien Province in southeastern China. Andrews first learned of the creature in letters from a famous missionary and big-game hunter, Harry Caldwell. A native of Tennessee, Caldwell operated a group of Methodist missions in Fukien. He traveled about the countryside "with a Bible in one hand and a rifle in the other," remarked Andrews, "[pursuing] his evangelical work among the Chinese, ridding their villages of man-eating tigers, while he poured into their ears the Eternal Truth." When plans for the Asiatic Expedition were announced, Caldwell invited Andrews to Fukien to join him in tracking down the animal. Hoping to secure a rare specimen for the Museum (Andrews correctly surmised that the "blue" tiger was a melanistic variation of a tiger's usual yellow-and-black coloration), Andrews accepted the challenge to the delight of the press, which relentlessly exploited the story.

By the time Andrews and Yvette reached China, the political climate was more chaotic than ever. Yuan Shi-kai, unable to exterminate his enemies, had been forced to renounce the throne on March 22, 1916, one hundred days after declaring himself emperor. He then issued a mandate canceling the monarchy and restoring the republic, but these concessions failed to placate his adversaries, who demanded his removal from power. Hardly had Yuan relinquished the throne when he became desperately ill with a kidney disease. He died on June 6, leaving his country shattered and unable to head off the horrors of the warlord era that would tear it apart for almost twenty years.

After spending a week in Peking, the couple left for Shanghai by rail to purchase supplies for their inland journey. Next, they traveled southward to the mouth of the Min River and went by motor launch to Foochow. Bound for Harry Caldwell's mission in Yen-ping, Andrews, Yvette, and their servants, accompanied by thirty-three coolies carrying sixteen hundred pounds of baggage, glided slowly up the Min River by launch and sampans. Nine days later, they reached the remote village of Chang-hu-fan "where Mr. Caldwell stood on the shore," wrote Yvette, "waving his hat to us amid scores of dirty little children and the explosion of firecrackers." A tall, athletic man, whom Andrews and

Yvette instantly liked, Caldwell joined the expedition, which continued on to Yen-ping, located in the north-central part of Fukien Province. Here, Caldwell and his family operated a Methodist mission, with a school and hospital, situated inside a walled compound on a hill just outside the town. Normally a peaceful farming community, Yen-ping was still engulfed in the hostilities that had erupted in the southern provinces as a consequence of Yuan's seizure of the throne and his sudden death. Heavy fighting between revolutionaries and northern troops sent from Peking had left Yen-ping littered with dead and wounded. Amid panic and bullets "[humming] in the air like angry bees," Andrews and Yvette, "with rude crosses of red cloth pinned to our white shirts," rushed about with four Chinese stretcher-bearers, collected wounded soldiers and civilians, and carried them to the mission hospital. After several days of gunfire, looting, and beheadings, the fighting was quelled by the arrival of a large contingent of northern soldiers from Foochow.

Once calm was restored, Andrews' party, with Caldwell in tow, went in search of the blue tiger. Ever since Caldwell first learned of its existence in 1910, it had always been sighted in the vicinity of Futsing, about thirty miles southwest of Foochow, in an area of forested mountains, cultivated valleys, and ravines choked with sword grass. Often, the tiger was reported almost simultaneously in places several miles apart. "So mysterious were its movements," Andrews commented, "that the Chinese declared it was a spirit of the devil."

But like his pursuit of the "Great Invisible" at Musan in Korea, this hunt, too, ended in frustration. Operating from a base camp at Futsing, Andrews and Caldwell chased the tiger with dogged determination, wearily trudging from one village to another as the animal wandered about killing goats, pigs, and dogs. Time and again they nearly bagged the elusive beast, but it always escaped at the last moment as if truly possessed, as the Chinese insisted, by "a demonic guardian spirit." After five weeks, the hunt was abandoned.

Late in July, Edmund Heller joined the expedition at Futsing. Heller was an expert zoologist who had collected in many parts of the world, including trips to Africa with his friend Carl Akeley and ex-president Theodore Roosevelt. With Heller's arrival, the Asiatic Zoological Expedition began in earnest. Sailing by way of Hainan Island, the travelers set out for Haiphong. From there, they went to Hanoi, the capital of what was still the French colony of Vietnam. After reorganizing approximately four thousand pounds of baggage, everything was put aboard a train, along with the servants and mule drivers or *mafus*, for a slow trip through fever-stricken jungles to Yunnan Fu, the province's capital city. On August 6, a caravan of horses and mules carrying half the supplies was dispatched to Ta-li Fu, an ancient trading center nestled at the foot of snow-covered mountains. Three days later, the explorers, carrying the rest of their provisions, followed the caravan on horseback. At one point along the way, bandits, striking just ahead of them, attacked a procession belonging to a local mandarin, made off with $5,000 worth of jade and gold dust, and sent the *mafus* and pack animals running for their lives. Andrews derived little comfort from the fact that his party was accompanied by a military escort provided by the foreign office in Yunnan Fu. "The first day out we had four [guards]," Andrews recalled, "all armed with umbrellas." The next day, they were replaced by another group equipped with rifles of an 1872 vintage, "but their cartridges were seldom of the same caliber as the rifles and in most cases the ubiquitous umbrella was their only weapon."

From Ta-li Fu, the expedition made its way slowly toward the northwest, past spectacular eighteen-thousand-foot mountains teeming with animals. By mid-November, the travelers had reached the Yangtze River and explored a maze of steep canyons, forests, and bamboo thickets, eventually reaching the Tibetan border and venturing as far west as the Mekong River. On January 13, 1917, they began a difficult journey from the Mekong more than a hundred miles southward to the Nam-Ting Valley, pausing frequently to collect wildlife. Continuing westward, the expedition finally reached the Salween River, which they crossed by ferry before striking out toward the end of April for a major trading center called Teng-yueh.

On June 1, with forty-one cases of specimens loaded onto thirty mules, the caravan left Teng-yueh headed for Burma. A week later, the exhausted travelers, after descending from the highlands into dense jungles, arrived in Bahamo on the Irrawaddy River. Altogether, they had spent nine months in Yunnan Province alone, covered over two thousand miles, camped at 108 different locations, and journeyed from altitudes of fourteen hundred to fifteen thousand feet above sea level. They had visited and photographed many indigenous non-Chinese peoples, such as the Lolos, Shans, Mosos, Palaungs, Lisos, and Kachines, groups about which virtually nothing was known in the outside world. And most important from Andrews' viewpoint, they encountered rich faunal zones that varied from semitropical to alpine and yielded a huge assortment of wildlife ranging from tiny rodents to exotic species such as gorals, muntjac deer, serows, civets, monkeys, gibbons, and leopards. When the expedition ended in Bahamo, Andrews noted that his forty-one packing cases contained the following inventory:

```
 2,100 mammals
   800 birds
   200 reptiles and batrachians
   200 skeletons and formalin preparations for anatomical
       study
   150 Paget color plates
   500 photographic negatives
10,000 feet of motion picture film
```

At Teng-yueh, news reached Andrews that the United States and Germany had declared war, an event that seriously hampered the shipment of the collections to the Museum, since they could no longer be transported across the Atlantic because of the disruption of commercial routes. Andrews had planned to return to New York by sailing from Burma through the Suez Canal and the Mediterranean. As this was now impossible, the only alternative was to travel from Bahamo down the Irrawaddy River to Rangoon, cross the Bay of Bengal to Calcutta, and traverse the breadth of India to Bombay. "A hot impatience robbed me of pleasure in the wondrous sights of India," Andrews complained, "and three weeks waiting . . . in the stifling heat at the Taj Mahal Hotel in Bombay drove me nearly wild." Finally, he was able to secure passage on a ship bound for Singapore, then retrace his route back to Hong Kong, Japan, and across the Pacific. Thus ended a remarkable trip, chronicled by Andrews in an enthralling book, *Camps and Trails in China*, which was published in 1919 and included six chapters written by Yvette, along with a selection of her photographs.

Andrews and Yvette arrived in New York on October 1, 1917, well satisfied with their accomplishments, besieged as usual by reporters, and filled with nostalgic reflections on their adventures. "Once more we are undefinable units in a vast work-a-day world," Andrews wrote, "bound by the iron chains of convention to the customs of civilized men and things." Still, there were practical considerations that made it necessary for them to leave China: the uncertainties of World War I, and the fact that Yvette was seven months-pregnant.

CHAPTER 6

On December 26, 1917, Yvette gave birth to a son, George Borup Andrews, named after his uncle, whose drowning five years earlier was still deeply felt by Yvette. Once his wife and son were settled for the duration of the war in their Lawrence Park house in Bronxville, Andrews went to Washington to call on an acquaintance he had met in New York, Newton D. Baker, the secretary of war. He arranged the meeting to volunteer his services in France, "wishing desperately," he declared, "to be part of the action." Baker, however, declined his request, arguing instead that Andrews could be more useful to the war effort with the army's intelligence corps in Washington, precisely the sort of desk job he wanted to avoid.

But as so often happened in Andrews' life, the unexpected intervened. While eating lunch later that day at the Cosmos Club, "sitting disconsolately at a small table by myself," he recalled, he was spotted by a friend, Charles Sheldon, a wealthy sportsman and big-game hunter who was serving with naval intelligence. Before the afternoon ended, Sheldon had recruited Andrews to return to the Far East as a full-fledged spy for the navy.

On June 10, 1918, Andrews took his oath of allegiance in Washington, and signed an agreement setting forth the terms of his obligations and salary—$4 a day per diem, another $4 daily "in lieu of subsistence," plus all travel and incidental expenses. A week later, Andrews was on his way back to Peking, where he established his base of operations under cover of carrying on zoological collecting for the Museum. He had been given the code name "Reynolds" and a letter of introduction to a Commander I. V. Gillis, his superior in China. It was unsigned, written in invisible ink, and concealed between the lines of a falsified letter supposedly sent by an "Allan Fraser" to "J. W. Rankin," neither of whom existed.

Andrews arrived in Peking to find it "a city of intrigue." Every new-comer to the clubs and hotels frequented by the foreign community was rumored to be a secret agent. On August 14, 1917, China had en-tered the war on the side of the Allies, though its contribution was lim-ited mainly to labor battalions sent to work behind the lines in Europe, the Middle East, and Africa. Within China itself, Andrews observed, the newly formed "republic" was a joke. Warlords controlled every province, each with his private army of conscripts and bandits. Armies were bought and sold like merchandise, and graft and corruption were epidemic, fu-eled by a thriving opium trade. Local conflicts continually erupted, usually with more bluster than actual killing. As one Chinese officer, Admiral Tsai Ting-kan, jested when someone commiserated with him about the country's civil wars, "You forget how very civil they are." And Andrews pointed out that during an abortive attempt to capture Peking, a newspaper reported, "There were ninety thousand rounds of ammu-nition fired today and one man was killed. He was a peanut vendor."

Throughout his lifetime, Andrews cloaked his activities for naval intelligence in an aura of secrecy. In *Under a Lucky Star*, he wrote, "It is not permitted even now [1943], to tell what part I played in the war." And in his 1929 memoir, *Ends of the Earth*, he was equally cryptic. He revealed only that his travels carried him to several provinces in China, through sections of Manchuria, and by horseback into Siberia's thickly forested wilderness. Most important from the standpoint of Andrews' future career, he also made two trips by automobile from Kalgan on China's northwestern frontier across the plains of Mongolia to its capi-tal city of Urga.

Recently declassified documents confirm that Andrews was one of many civilian informants, operating under various guises, whose job it was to gather data on a wide range of subjects for use in formulating American policies in eastern Asia. Andrews, for example, filed reports dealing with, among other things, political conditions in China, communication and rail facilities, troop movements, industrial output, armaments, shipping and ports, and evidence of foreign intervention — particularly Japanese — in China and Manchuria. While traveling in

Mongolia, Andrews' observations focused on internal affairs, military capabilities, and potential effects of the Bolshevik Revolution, then raging inside Russia, an event that would exert a profound influence on Mongolia's political fortunes.

Aside from the covert reasons for his forays into Mongolia, these journeys gave Andrews an opportunity to evaluate this part of central Asia for future scientific exploration. Not only would he conclude that it offered "undreamed-of possibilities for testing Osborn's ideas regarding mammalian evolution," but he also felt a compelling affinity with the land and its fiercely independent nomads. From the moment he set foot in Mongolia on a reconnaissance trip in August 1918, he was smitten by the country's overpowering landscape. After passing through the Great Wall at Kalgan (now renamed Ch'ang-Chia-k'ou), on his way to Urga, he was enveloped by astonishing vistas. Behind him, gently rolling hills stretched away to where the Shansi Mountains met the horizon. Ahead, the terrain was "cut and slashed by the knives of wind and frost and rain," Andrews wrote, "and lay in a chaotic mass of gaping wounds—cañons, ravines, and gullies, painted in rainbow colors, crossing and cutting one another at fantastic angles." A few miles farther on, at the summit of a pass opening onto the Mongolian plateau, was a seemingly infinite expanse of undulating plains that swept to the west and south toward the Gobi Desert.

Beneath its awesome beauty and mysticism, Mongolia could be forbidding and cruel; its environment was harsh and its history drenched in blood, as Andrews would learn soon enough. Still, like so many travelers before him, he was instantly seduced by the country's hypnotic lure, and he described his first impression as if through the eyes of a heedless romantic:

> Never again will I have such a feeling as Mongolia gave me. The broad sweeps of dun colored gravel merging into a vague horizon; the ancient trails once traveled by Genghis Khan's wild raiders; the violent contrast of motor cars beside majestic camels fresh from the marching sands of the Gobi! All this thrilled me to the core. I had found my country. The one I was born to know and love.

During Andrews' intelligence missions, his only brush with serious danger occurred on his first trip to Urga. Accompanied by another American, Charles Coltman, a resident of Kalgan who regularly drove to Urga on business, Andrews was at the wheel when five men attacked them near Ude, the border crossing between Inner and Outer Mongolia. As the assailants suddenly opened fire from a ridge overlooking the road, a hail of bullets hit the car, one of which shattered part of the steering wheel just as Andrews was reaching into the back seat for his rifle. Swerving at full throttle behind an outcropping of rock, Andrews drove the car into a dry streambed, where it became mired. "Leaving the motor running," he recounted, "we climbed the rocks and peered over the top. The five men stood in plain sight, discussing what to do next. They were dressed in Mongol robes, but that proved nothing. They might have been Russians. Because their shooting was so accurate, I was certain they weren't Chinese, who are notoriously bad marksmen."

As the would-be murderers scrambled down the escarpment and headed toward the stranded car, "evidently bent on finishing off what they had started," Andrews continued, Coltman took aim at one man silhouetted against the sky. Andrews lined his sights on another walking in front. They fired simultaneously, apparently killing both men, who slumped to the ground and lay motionless, while the other three fled into the rocks for safety.

News of the Armistice on November 11, 1918, reached Andrews while he was in Shantung Province. Within a few days, he returned to Peking, delighted to find its foreign residents "still *en fête* [with] an endless round of dinners, dances, and celebrations." Although he repeatedly proclaimed his "bitter disappointment" over not having been sent to France, Andrews clearly savored the adventurous aspects of his stint as a spy in China and Mongolia—not to mention its beneficial impact on his future career. By now, he had acquired considerable knowledge of the geography and logistical problems of both countries, and even picked up a smattering of the Mandarin language widely spoken in China. He had cultivated an extensive circle of friends and highly placed contacts that included influential Chinese officials, Mongolian dignitaries, businessmen and traders, missionaries, scientists, military officers, and diplomats attached to the United States and European legations in Peking. He also gained valuable insights into Chinese and Mongolian customs, history, and political machinations (to the extent that the latter could ever be comprehended by Westerners); and, socially, he had become a "regular" among the international mélange of expatriates and travelers who constituted Peking's foreign community.

Most important, Andrews' clandestine journeys into Mongolia had inspired plans for what would become known as the Second Asiatic Zoological Expedition. Andrews envisioned it as the next phase of the concept he had presented to Osborn in 1915, and its objective was to collect animals and birds in northern Mongolia. Remaining in Peking, Andrews rented a secluded house in a walled compound at 36 Wu Liang Tajen Hutung, near the city's legation district. Yvette and George had arrived toward the end of September, and, along with household servants, a Swiss nurse was engaged to care for George, who was now nine months old. Andrews then set about arranging for his new expedition entirely by cables and letters to the Museum. Osborn gave his unconditional approval, agreeing to provide half of the project's relatively modest budget, estimated by Andrews at $7,500. The balance was donated by Andrews' friends, the German-born textile magnate and president of the Bear Mill Manufacturing Company in New York, Charles L. Bernheimer and his wife, who had contributed to the Yunnan expedition three years earlier.

Andrews' service with naval intelligence was not terminated until April of 1919, and perhaps only then because of an embarrassing episode involving Yvette. In December 1918, she had sent letters to relatives in the United States in which she deliberately revealed Andrews' role as a spy. Writing to an aunt in Ossining, for example, she blatantly stated:

> Now that the war is over I can tell you that the heading of this writing paper [which bore the imprint Second Asiatic Zoological Expedition of the American Museum of Natural History] is mere camouflage and that Roy really has been sent out by the Government on a Secret Mission to China.
>
> It was hard not to tell the truth to you, but I know you'll understand and be glad with me that Roy could be of service at this time.

Just at present there are many very brilliant men out here on different missions. . . . One has to be on one's tip toes all the time and it's piles of fun.

Why Yvette wrote such compromising letters, and whether Andrews knew about them in advance, is unclear. Whatever the case, her letters were intercepted by censors and touched off a barrage of top-secret telegrams between high-ranking officials. On February 6, the American minister in Peking, Paul S. Reinish, who had assisted Andrews in securing visas for his Yunnan expedition in 1916, was contacted by the acting secretary of state:

It is indicated by reports through censorship that information is being circulated by Mrs. Roy Chapman Andrews to the effect that her husband is on a secret Government mission. Whether founded on fact or not, such publicity is unfortunate. Report what Andrews' mission and connections are after discreet investigation.

And there the incident seems to have ended, or at least no further documentation has surfaced regarding its outcome. We know only that Andrews, saved perhaps by his persuasive charm or a plea of ignorance concerning his wife's letters, was apparently not held accountable for the matter. He retired from naval intelligence in April, possibly because of Yvette's indiscretion, but it is unlikely that he ever received a reprimand. Indeed, by the spring of 1920, he was again supplying military information on China and Mongolia to the government—only then it was to the chief of staff's office at the War Department and in an unofficial capacity.

Obviously, the ruckus caused by Yvette's bad judgment did not deter plans for Andrews' Second Asiatic Expedition. By February 1919, most of the equipment and provisions had been sent to Urga by caravan. Andrews' friend Charles Coltman, who operated trading stations in Kalgan and Urga, secured automobiles for the journey to the Mongolian capital, where the expedition would actually get under way. On May 17, three cars transporting baggage, Yvette's camera gear, and a high-spirited group of travelers began the arduous drive, following

Mongolia's only "road"—a well-worn gravel caravan trail that extended for roughly seven hundred miles between Kalgan and Urga. In addition to Andrews and Yvette, the party included Charles Coltman and his wife; two Chinese taxidermists, Chen and Kang, engaged by Andrews along with a cook known as Wu; a soldier named Owen, hired by Coltman to drive one of the cars; and Mr. and Mrs. Ted MacCallie, friends of the Coltmans.

Aided by Wu's excellent cooking, plentiful game, and an ample supply of liquor, everyone's mood was jubilant. Andrews, thrilled to be back in the field and reunited with Yvette, described the trip as *"une belle excursion,"* and he jovially dubbed his traveling companions as "the Grouchless Gang." Every few miles, Andrews, Coltman, and MacCallie stopped to shoot gophers, fat yellow marmots, rabbits, wolves, and fleet-footed antelope for the collection. And there were flocks of geese, mallards, sheldrakes, teals, and cranes, which fell in abundance before their shotguns, the geese and ducks providing savory meals when cooked in the makeshift oven Wu improvised from a gasoline can. Entranced by her first glimpse of Mongolia, Yvette photographed wildlife, nomad encampments, landscapes, passing caravans, and lamaseries inhabited by saffron-robed priests. Yet nothing, not even Andrews' vivid descriptions, had prepared Yvette for her first glimpse of Urga—the "City of the Living God"—which they reached after driving for nearly a week.

Founded sometime around 1649 as a trading center at the confluence of several major caravan routes, Urga retained an untamed, frontierlike quality that had long fascinated travelers. "The world has other sacred cities," wrote Andrews, "but none like this. It is a relic of medieval times overlaid with a veneer of twentieth-century civilization; a city of violent contrasts and glaring anachronisms." He marveled at the spectacle of nomadic tribesmen trotting along on their small, powerful horses, indifferent to motor cars that roared past; the long sauntering lines of heavily loaded camels "fresh from the vast, lone spaces of the Gobi Desert"; crowds of Lamaist monks in brilliant red and yellow gowns; and swarthy Mongol women clad in elaborate costumes with fantastic bejeweled headdresses "staring wonderingly at the latest fashions of their Russian sisters." In Urga's dusty streets and crowded shops, Mongols in knee-length robes tied with brightly colored sashes mingled with Chinese, Manchu Tartars, camel drivers from Turkestan, and a scattering of Russians, Scandinavians, and other Europeans. Now and then, Andrews recalled, "a resplendent group of horsemen in flaming red, purple, and gold robes, wearing pointed yellow hats with peacock feathers, dashed down the street at full gallop. . . . In its kaleidoscopic mass of life and color, the city was like a great pageant on the stage of a theater."

Situated in northeastern Mongolia, roughly 160 miles below the Russian border, Urga sprawled along the banks of the Tola (Tuul) River. Surrounded by rolling, grassy hills, it lay in a valley at the foot of the Bogdo-Ola (God's Mountain), whose thickly wooded slopes rose to an altitude of eleven thousand feet above sea level, shutting off Urga from the sweeping plains to the south that eventually melted into the Gobi. Andrews estimated Urga's population at about twenty-five thousand people, some fifteen thousand of which, he said, were priests.

Approaching the city from the south, along the trail from Kalgan, one entered the Chinese quarter, or Mai-ma-cheng. For the next two miles, the narrow street was bordered by gaudily painted Russian cottages dominated by Russia's consulate, a huge red building that impressed Andrews as a monstrosity of "surpassing ugliness." Farther on, the road opened onto a wide square, an indescribable mixture of Russian, Mongolian, Tibetan, and Chinese architecture. "Palisaded compounds," Andrews recalled, "gay with fluttering prayer flags, ornate [Russian] houses, felt-covered yurts, and Chinese shops mingle in a dizzying chaos." Every house and shop was protected by stockades of unpeeled timbers, giving the city a rustic quality Andrews compared to "an American frontier outpost of the Indian fighting days." Dominating a hill near the main square loomed the resplendent Gandan Lamasery, the city's principal religious complex, which drew pilgrims from all over Mongolia who came to prostrate themselves before sacred images, spin prayer wheels, and chant barely audible mantras. And scattered among the houses and shops were ubiquitous shrines, golden-domed temples, and lamaseries—some incredibly ornate—which embodied elements of Tibetan and Chinese architecture.

At the eastern end of Urga's central street stood the Ministry of Foreign Affairs, consisting of several wooden structures encircled by a stockade. Nearby, inside an enormous compound jammed with loaded carts and camel caravans, was a crude building that served as the customs house, although most of its business was conducted in a large yurt equipped with desks, a metal filing cabinet, and a telephone. Just beyond the customs house was what Andrews denounced as "one of the most horrible prisons in the world." Shielded from view by a double palisaded wall was a building filled with stacks of wooden boxes four feet long by two and a half feet high. Seething with revulsion, Andrews discovered that these "coffins" were actually the prisoners' cells in which "the poor wretches [some with] heavy chains about their necks and both hands manacled together . . . [could] neither sit erect nor lie at full length." The agony of their cramped position, Andrews raged, "is beyond the power to describe. . . . Sometimes they lose the use of their limbs, which shrink and shrivel away."

Despite its rough-and-tumble frontier atmosphere, Urga was the holiest city in the country, the dwelling place of Mongolia's most venerated spiritual leader, known to his subjects as the Hutukhtu, Living Buddha, or the Bogdo Gegen. While he ranked below the Dalai Lama in Tibet, who stood at the pinnacle of the Buddhist hierarchy, the Hutukhtu was recognized as a legitimate reincarnation of Buddha, and therefore wielded enormous religious and secular power. Although he was devoted to his wife, Ekh Dagin Dondogdulam, whom he had married in violation of Lamaist monks' traditional vows of celibacy, it was rumored that during former years, the Hutukhtu enjoyed "the pleasures of the flesh," leaving his "heaven," as Andrews put it, "to revel with convivial foreigners in Urga." Now old, infirm, and nearly blind, the Bogdo Gegen was living out his final days in three palaces near the Tola River surrounded by his cherished collection of European and American furniture, antiques, mechanical gadgets (many ordered from a Sears, Roebuck catalog), automobiles, and mounted animals from around the world. Years before, the forest that cloaked the Bogdo-Ola, or God's Mountain, south of Urga, had been set aside as the Hutukhtu's private game preserve, guarded by as many as two thousand lamas who did not hesitate to kill intruders attempting to poach within its sacred boundaries.

While in Urga, Charles Coltman introduced Andrews and Yvette to a Swede generally recognized as the most respected foreigner in Mongolia, one F. A. (Franz Augustus) Larsen, who was widely known as "Larsen of Mongolia." Born in Stockholm in 1870, he had come to Mongolia at the age of twenty-one and immediately fallen under its spell. At first, Larsen and his wife served as missionaries with the American Bible Society in Inner and Outer Mongolia, but he soon abandoned the frustrating task of spreading the Gospel among an unreceptive populace in favor of indulging his passion for raising and selling horses. He built a house and breeding facility near Kalgan and maintained another compound in Urga.

Few outsiders ever acquired a deeper understanding of Mongolia's customs, history, religious beliefs, languages, geography, and wildlife. He enjoyed intimate friendships with the country's premier and ruling khans, and because he was conversant in Mandarin, the Hutukhtu often sent him as an emissary to Peking to resolve political difficulties with the Chinese, a service for which the Living Buddha awarded him an honorary title as the "Duke of Mongolia." By the time Andrews met him in the summer of 1919, Larsen was still dealing in horses and camels, though for several years he had been in charge of the Urga branch of a European-owned trading operation known as the Anderson, Meyer Company, which had its main office in Kalgan.

Larsen's assistance proved invaluable in launching Andrews' expedition. He helped outfit the party with scarce supplies, arranged to borrow two heavy-duty carts and purchased a third, along with horses to pull them. He hired a young lama who spoke enough Chinese and pidgin English to act as interpreter and loaned him a pony. From a Mongol prince named Tze Tze, he acquired sturdy riding horses for Andrews and Yvette. The gelding Larsen gave to Andrews, Kublai Khan, proved to be so spirited, surefooted, and "utterly lovable" that Andrews took him back to Peking and rode him for years. Yvette's chestnut stallion, on the other hand, was "a tricky beast," Andrews fumed, "whom I could have shot with pleasure."

After ten days in Urga, the Second Asiatic Zoological Expedition departed. "Our arrival in Urga was in the approved manner of the twentieth century," Andrews recorded in his journal. "We came in motor cars with much odor of gasoline and noise of horns. When we left the sacred city we dropped back seven hundred years and went as the Mongols traveled—on horseback, pulling overflowing carts along rough, deeply scarred trails." Andrews and Yvette led the way, followed by the sullen lama interpreter wrapped in a yellow robe, while the cook, Wu, and the two Chinese taxidermists, Kang and Chen, bounced along in the carts, clinging to the piles of equipment "with an air of resignation and despondency."

Their destination was a district three hundred miles southwest of Urga. According to Andrews' informants, it was an area of windswept plains, reportedly teeming with antelope and small mammals, which gave way to jagged mountains inhabited by ibex and bighorn sheep. But it was soon obvious that the area had been overhunted by nomads. Except for geese, demoiselle cranes, ducks, skylarks, and a few rodents, there was virtually no wildlife on the plains, and Andrews was not equipped to extend his search into the mountains.

Returning to Urga on June 16, Andrews decided to revisit the eastern plains through which they had previously traveled on the way from Kalgan. Aside from antelope and marmots, this region abounded with a variety of other animals, and the lines of traps set out every night yielded a plentiful bounty, keeping the Chinese taxidermists busy throughout the day preparing specimens. Yvette labored diligently at photographing the nomads who continually visited the camp to invite the explorers to their yurts, bringing gifts of blue silk scarves, dried mutton, goat cheese, rancid butter (considered a delicacy), and the sour, intoxicating kumiss—or *airaq,* as it is also known—made from fermented mare's milk. In exchange, they received cigarettes, empty tin cans, or bars of colored soap.

Having spent two months on the eastern plains, the expedition journeyed north into the high country near the Russian border. Andrews described it as "a paradise blanketed with lush grass and eerie silent woodlands. . . . Never had we seen such brilliant wildflowers . . . acre after acre of bluebells, forget-me-nots, daisies, buttercups, yellow roses, and cowslips." Overcome by the grandeur of the landscape, Andrews decried its inevitable destruction by logging, just as automobiles, he felt, would despoil the plains and desert:

It is a country of exceeding beauty. The dark forests of spruce, larch, and pine, broken now and then by a grove of poplars or silvery birch, the beautiful valleys and the rounded mountain summits, are as wild and pristine as nature made them and seem untouched by the devastating hand of man. I wonder how long they will remain inviolate! Certainly not many years after the Gobi Desert has been crossed by lines of steel and railroad sheds have replaced the gold roofed temples of sacred Urga. I hope I shall not have to see it then. . . . To stand on the summit of a mountain as I did last night and gaze across the miles and miles of unbroken green, it would seem that here there was an inexhaustible supply of wood. But no more pernicious term was ever coined than "inexhaustible supply!" There is only so much water in the ocean and every bucket full that is removed makes it so much less.

By the first of September, a hint of fall was beginning to permeate the night air, causing Andrews' entourage to make its way back to Urga to arrange for the return trip to Peking. On October 1, the zoological collection was shipped out by camel caravan. Kublai Khan, Andrews' horse, plodded along under the watchful eyes of the drivers, while Andrews, Yvette, the taxidermists, and the cook drove in cars operated by the Chinese government. Although small in scale, the expedition was returning with over 1,000 skins of mammals and birds, almost 250 photographs, motion picture films, and notebooks crammed with observations on Mongolia's ethnography, history, religious practices, geography, wildlife, vegetation, and climate. "I had learned much about the country and its ways," Andrews commented, "talked with wandering Mongols about the far western desert, studied the physical problems of transport and maintenance in the arid reaches of the Gobi. . . . All I learned made me more certain that this was the chosen spot [for further explorations] in Central Asia; the place where I could stake all to lose or win on a single play."

After returning to Peking and depositing Yvette in their rented house with George, his Swiss nurse, and the servants, Andrews, inexhaustible as ever, was off on another trip. Summoning his tiger-hunting friend in Fukien, Harry Caldwell, they set out for the Shansi Mountains to secure specimens of argali sheep—the regal, mythical-looking creatures with heavy curved horns like battering rams that range from Mongolia's Altai Mountains to the Shansi peaks of northern China. Yvette was in "open rebellion" when Andrews refused to take her along, but the proliferation of bandits in the Shansi Mountains was so menacing that the area had been closed to hunters since 1915.

In a reckless display of bravado, Andrews and Caldwell ignored official warnings, organized an elaborate procession of native hunters, two Chinese taxidermists, *mafus*, and supply carts drawn by donkeys, and marched boldly into the forbidden recesses of the Shansi Mountains, utterly disregarding the threat of bandits. Reappearing after several weeks of hunting amid thickly forested canyons, Andrews and Caldwell brought out half a dozen magnificent examples of argali sheep, including one of the largest rams ever recorded. As a bonus, they had also killed wapitis, roebucks, gorals, and boars, and the taxidermists had trapped numerous small mammals and rare birds.

When the results of the Second Asiatic Zoological Expedition were finally cataloged, it had added fifteen hundred mammals and birds to the Museum's collections, "all from a region virtually new to science," Andrews pointed out with satisfaction. In addition, he contributed a series of articles on the journey to *Harper's Magazine*, *Natural History*, and other publications. "But the really vital thing that emerged from the trip," Andrews declared, ". . . was the awareness of Mongolia as a theater of work for the massive expedition that was taking shape in my imagination."

CHAPTER 7

Until this point, Andrews' ventures had been as much quests for self-discovery as for scientific knowledge. One of the driving compulsions underlying his remarkable achievements thus far was an attempt to find a clear-cut direction for his mercurial career. By the time he returned from China with Yvette and George in February of 1920, however, all doubts regarding the future course of his life had been erased. Andrews knew precisely what he wanted to accomplish in the years ahead; and it was this newly found resolve that brought about the genesis of the Central Asiatic Expeditions—the venture that would ultimately define his identity, bring him dazzling fame, and grant his lifelong wish: a place among the elite of twentieth-century explorers.

By now, the "great expedition of [Andrews'] dreams" far exceeded the plan he had presented to Osborn five years earlier. Its basic premise remained unchanged: to seek proof that Asia, Osborn's "Mother of Continents," was the origin point for most of the Northern Hemisphere's mammals, including mankind's ancestors. But Andrews had long since resolved that the scope of his future explorations should be expanded to encompass a multidisciplinary approach, one in which specialists representing a variety of fields could join together to study specific problems.

Three days after returning to New York in 1920, Andrews arranged a lunch with Osborn at the Museum. Aware that the professor loathed discussing business while eating, Andrews chatted idly about Museum affairs, family matters, and his recent travels in Mongolia and China. Osborn bided his time until coffee was served and the two men were leisurely smoking. He then gave Andrews the anxiously awaited cue: "Now let's have it, Roy," he said. "It's another expedition I suppose."

Andrews instantly unleashed a veritable torrent of information, everything he had been mulling over for months. "Talking," he wrote, "as I have never talked before," he elaborated on his reasons for selecting Mongolia—particularly the Gobi—as a likely place to test Osborn's theory of mammalian evolution and search for human fossils.

Andrews pointed out that his previous journeys through Yunnan, Mongolia, and northern China had focused solely on collecting zoological specimens for comparative study, with no attempt to gather the sort of paleontological and geological data necessary to trace the evolution of Asian mammals. Now Andrews' sense of mission was considerably grander. "We should try," he told Osborn, "to reconstruct the whole past history of the Central Asian plateau—its geology, fossils, [ancient] climate, and vegetation. We've got to collect its living mammals, birds, fish, reptiles, insects, and plants and map the unexplored parts of the Gobi. It must be a thorough job; the biggest land expedition ever to leave the United States."

Next he laid out a radically unconventional proposal to scour the countryside using automobiles supported by a camel caravan. "It will act exactly like the supply ship to a fleet at sea," he assured Osborn. Andrews had concluded from his own experience and information reported by nomads and caravaneers that Mongolia's gravel-covered surface, sporadically broken by mountains, rocky outcroppings, badlands, and sand dunes, would allow him to utilize motorcars to cover great distances rapidly and with relative mobility, assuming that the supply caravan loaded with gasoline, oil, spare parts for the cars, and extra food made it to each rendezvous point on schedule, without falling victim to bandits, drought, or insufficient grazing for the camels. Andrews cited the fact that camels averaged only ten or fifteen miles a day, whereas automobiles could travel a hundred. "If all went as expected," he wrote, ". . . we could do ten years' work in five months."

Another key aspect of his operation, Andrews explained, was the idea of dividing the expedition into three or four self-contained units. Each would consist of one or two cars carrying scientists, their assistants, and enough supplies and fuel so that every unit could work independently of the base camp for at least two weeks, thereby enabling several areas to be investigated simultaneously.

Osborn raised the point that no fossils had ever been reported from Mongolia, except for a single tooth of an extinct rhinoceros picked up along a caravan trail in 1894 by the Russian explorer V. A. Obrechev. Most scientists believed that even this scant bit of evidence had been dropped by traders, since fossils, or "dragon bones," as they were called in China, brought high prices in apothecary shops where they were sold as medicines and aphrodisiacs. But Andrews contended that previous explorers in Mongolia had never looked systematically for fossils. Moreover, they had relied on camels as their mode of transportation, which, when compared to automobiles, limited the range of their explorations, often forcing them to follow well-traveled caravan routes. Andrews further insisted that Mongolia's exposed terrain, deeply eroded in places and devoid of vegetation, made it easier to study its geology and search for fossils by means of modern scientific methods, which had never been applied in the region.

By the time Andrews ended his presentation, Osborn was persuaded. It was exactly the kind of undertaking that fired his entrepreneurial instincts: audacious, innovative, and sweeping in scientific scope. Nor was Osborn blind to the incalculable publicity value to the Museum of such a venturesome enterprise, or to the fact that Andrews—with his inexhaustible drive, salesmanship, talent for organization, raw courage, and experience gained from his prior explorations in Asia—was the right man to lead it.

Nothing comparable had ever been attempted in the history of exploration. From a scientific standpoint, it was a revolutionary concept; logistically, it posed a monumental challenge. A plan of such complexity would also involve a staggering amount of money, which the Museum could not provide solely from its own resources. Only an enormous infusion of outside support could possibly underwrite an expedition of such magnitude.

Andrews estimated that the project would require a series of expeditions over a five-year period at a cost of $250,000 (equivalent to about twenty times that much today). When Osborn asked how he proposed to get the money, Andrews' reply was unequivocal: he would make it a "society expedition with a capital S." Wealthy New Yorkers, he believed, tended to follow a leader. If he could start by gaining the support of a financial titan, like J. P. Morgan, Jr., for instance, "everyone will think it a 'must' for the current season. Have you contributed to the Roy Chapman Andrews expedition? If not, you're not in society. That's the idea."

Firmly convinced that this strategy would work, Andrews began a fund-raising odyssey that few, if any, explorers had ever attempted. Osborn, of course, was ideally positioned to help set Andrews' campaign in motion. Social events linked to philanthropic causes—charities to help the poor, hospitals, cultural institutions, and religious organizations like the YMCA—had long been de rigueur among his moneyed circles. Osborn's ties to New York's elite extended over three generations. Both his grandfather and father had established close relationships with the city's merchandising and industrial giants. William H. Osborn, Henry's father, maintained business and social connections with Cornelius Vanderbilt, Morris K. Jesup, Jay Gould, and Jonathan Sturges, a prominent businessman and president of the New York State Chamber of Commerce, whose daughter Virginia was William's wife. E. H. Harriman (W. Averell's father) began his career as treasurer of William Osborn's railroad, the Illinois Central; and William belonged to many of the exclusive clubs frequented by the city's most affluent residents.

Apart from his family's alliances with millionaire friends and associates, Osborn's position as president of the Museum involved constant fund-raising efforts with socially obligatory overtones. It also brought him into contact with a group of powerful financial magnates that included Cleveland H. Dodge, Arthur Curtiss James, George F. Baker, Childs Frick, J. P. Morgan, Jr. (and his father, who died in 1913), Felix Warburg, George S. Bowdoin, W. Averell Harriman, William Rockefeller, and Theodore Roosevelt, Jr., all of whom were either trustees or had other involvements with the Museum. In short, Osborn enjoyed access to most of the city's civic-minded philanthropists, and opening the "right" doors for Andrews was not difficult.

Within a few days after his lunch with Andrews, Osborn had prevailed upon the Museum's director, Frederick A. Lucas, and the trustees to sanction the project and create a special fund that would contribute $5,000 a year toward the expedition's operating expenses.

Because Osborn agreed with Andrews that it would be highly advantageous to enlist J. P. Morgan, Jr., as his first sponsor—thereby setting a towering example for New York society to emulate—a meeting with the legendary financier was arranged at the Morgan Library on East Thirty-sixth Street. At nine o'clock on a windy January morning, Andrews, carrying a map of central Asia, was ushered into Morgan's sumptuous, art-filled office. As a trustee of the Museum since 1908, recruited by Osborn soon after he became president of the board, Morgan felt a deep affection for the institution his father had helped to establish in 1869 and subsequently enriched with collections and money.

Sensing Morgan's receptive frame of mind, Andrews unfolded his map and spread it on the desk. "There is always something exciting about a map," he later wrote, "and this was particularly true in those days when a lot of blank areas were still marked 'unexplored.' The entire Central Gobi was a white space with only a few thin lines waving uncertainly across it." Andrews launched into his pitch with unbridled enthusiasm, pausing occasionally to answer Morgan's questions. He elaborated on the impetus behind his proposed explorations—Osborn's theory concerning Asia's role in mammalian origins. He explained his revolutionary modus operandi: how a camel caravan loaded with supplies and fuel would be sent weeks in advance of the main party to drop provisions at predetermined points hundreds of miles into the desert; how his scientific staff and their assistants, traveling in automobiles, would follow early in the spring, spending five months each year probing every facet of Mongolia's natural history.

Writing in *Under a Lucky Star*, Andrews recounted that after fifteen minutes, Morgan's eyes were glowing with excitement. "It's a great plan," he declared. ". . . I'll gamble with you. How much money do you need?" Andrews' recollection of Morgan's response differs slightly in other published accounts of that landmark moment, but regardless of his exact words, there was no doubt about Morgan's enthusiasm: he pledged $50,000 toward the expedition's quarter-of-a-million-dollar budget! "Now go out and get the rest of it," he reportedly told Andrews, who could hardly believe what had just occurred. Morgan's gift was precisely the send-off he had wanted, and the amount exceeded all expectations.

Another individual on Andrews' list of prospects was the elusive

Andrews at the ages of
fourteen (ABOVE) and
twenty (BELOW).

ABOVE: Andrews at his headquarters
compound in Peking, 1925.

Andrews in Japanese dress at a hotel in Shimizu, Japan, 1910.

LEFT: A section of a California gray whale being hoisted onto the wharf at Ulsan, Korea, 1912.

LEFT: Andrews (with a mustache) dressed in Korean clothing, 1912.

BELOW: Andrews on the deck of the yacht *Adventuress* on an ill-fated trip to the Arctic Ocean to kill a bowhead whale, 1913.

Yvette Borup Andrews with her son George, ca. 1921.

Andrews and Yvette with animals collected in Yunnan Province, China, 1916.

The former Manchu palace in Peking that served as Andrews' home and the headquarters of the Central Asiatic Expeditions from 1921 to 1932.

The drawing room of the expeditions' headquarters in Peking.

Walter Granger, chief paleontologist and second-in-command of the Central Asiatic Expeditions.

Henry Fairfield Osborn, Mongolia, 1923. Osborn's theories regarding the evolution and dispersal of mammals, including the earliest humans, inspired Andrews' massive series of expeditions to the Gobi.

OVERLEAF: Andrews' fleet of automobiles—the first attempt to take vehicles across the Gobi's uncharted expanses. The third car flies the flag of the Explorers Club.

A part of the caravan used to supply Andrews' automobiles with gasoline, tires, extra parts, food, and other provisions. Depending upon the size of each of Andrews' five journeys into the Gobi, these support caravans varied from 50 to 125 camels, and traveled several weeks in advance of the automobiles.

RIGHT: Andrews and his staff engaged in the constant task of repacking supplies for transport by the caravan.

BELOW: Wooden crates being loaded with supplies and specimens for transport by caravan.

Staff of the 1923 expedition at Irdin Manha. *Second row, left to right:* Granger, Osborn, Andrews, Morris, Kaisen. *Top row, second from left:* Vance Johnson, Albert Johnson, Young, Olsen.

A Mongol inspecting one of the supply camels.

Luncheon in the mess tent emblazoned with Chinese symbols in gold on a dark blue background.
Shabarakh Usu, 1925.

BELOW: Expedition encampment in Inner Mongolia, 1928.

RIGHT: Tserin, the caravan leader in 1928 and 1930.

Mongol nomads listening to a phonograph. Invariably they were terrified when hearing it for the first time.

Andrews amputating a Mongol girl's badly infected finger. Tsagan Nor, 1922.

BELOW: Nomads, whom Andrews suspected were bandits, inspecting one of the expedition's Fulton trucks.

LEFT: Walter Granger with a pet red-billed chough.

BELOW: Nomads riding camels that are beginning to shed their heavy winter coats, which, when pulled off, supplied ideal packing material for fossils.

OVERLEAF: Supply caravan arriving at the Flaming Cliffs at Shabarakh Usu, 1925.

Andrews and George Olsen inspecting a nest of dinosaur eggs at the Flaming Cliffs, 1925.

A cluster of dinosaur eggs unearthed at the Flaming Cliffs. Originally they were believed to have been laid by *Protoceratops andrewsi*, but recent discoveries have disproved this assumption (see Epilogue).

Skull of a *Protoceratops andrewsi* found at the Flaming Cliffs.

Skeleton of *Protoceratops andrewsi* from the Flaming Cliffs. Now displayed in the American Museum of Natural History.

RIGHT: Skeleton of an *Oviraptor* found lying directly above a nest of dinosaur eggs at the Flaming Cliffs, leading to the mistaken conclusion that *Oviraptor* was preying on the eggs (see Epilogue).

Encasing fossils in burlap soaked in plaster of paris to prepare them for removal.

Granger supervising the packing of prepared fossils in crates, Shabarakh Usu, 1925.

Arthur Curtiss James, a trustee of the Museum who owned vast copper, silver, and gold mining interests along with a railroad empire that included the Western Pacific line. James ranked among the country's wealthiest magnates, and was once named by financial analysts as one of the fifty-nine men who "ruled" America. Unfortunately, Andrews' appointment with James was ill timed. He arrived at his Wall Street office to find the notoriously irritable tycoon fuming over a letter from Osborn requesting a donation to pay off the Museum's deficit. "There shouldn't be a deficit," he raged. "[Osborn] ought to spend his income and no more." After enduring James' ire, which was suddenly directed at Andrews after he, too, asked for money, Andrews reached for his hat and walking stick and started to leave. At that point James' anger subsided. But in *Under a Lucky Star*, he related that the meeting actually ended on a humorous note when James rather sheepishly apologized. "Now, Roy," he quotes James as saying with uncharacteristic humility, ". . . I can't let you go off to China being sore at me. Will ten thousand do you any good? . . . Teach me not to lose my temper. Always ought to pay for it when I get mad."

About this time, Andrews met Louis Froelich, the editor of the magazine *Asia*. Published by the American Asiatic Association, an organization founded in 1898 to promote public awareness of Asian affairs and "foster and safeguard American trade and commercial interests" in the region, the magazine had been a creation of Willard D. Straight, a young financier, diplomat, and sportsman with extensive Far Eastern experience. Seven years before his premature death from influenza in 1918, Straight had married Dorothy Payne Whitney, a daughter of the millionaire racehorse owner William C. Whitney. *Asia*, which Dorothy Straight continued to sponsor after her husband's death, enjoyed a large readership, especially among rich, well-traveled readers who regarded the Orient as an exotic destination for safaris and tours. Its contents ranged over the full gamut of Oriental history and culture—articles, for example, on Japanese art and architecture, China's political affairs, tiger hunting with Indian maharajas, and the social life of Peking's foreign colony.

Louis Froelich saw the possibility of a mutually beneficial link between *Asia* and Andrews' project, so he introduced the explorer to Dorothy Straight. Instantly charmed by Andrews, she endorsed

Froelich's idea of involving *Asia* in the expedition, and made a personal contribution of $5,000. Once the groundwork for such a partnership had been laid by Froelich, Andrews, Dorothy Straight, and Osborn, the executive committee of the American Asiatic Association approved the merger on June 14, 1920. *Asia* agreed to donate $4,000 annually toward the expedition's budget (its sponsorship eventually totaled $30,000) in exchange for a series of articles on the venture, plus footage for a motion picture to which the association would own the rights. A further stipulation stated that the designation "In Cooperation With the American Asiatic Association and *Asia* Magazine" was to be added to the expedition's official title.

Aside from helping to engineer the alliance with *Asia*, Andrews' meeting with Dorothy Straight proved valuable in other ways. A tall, elegant woman, renowned for her beauty, she had established herself as a pillar of New York society. She was also a person of diverse interests, with the financial resources to implement them. Two of her most significant achievements were founding the New School for Social Research and launching the *New Republic* magazine, which she started in 1914 with her husband, Willard. With Andrews' enterprise now high on her list of favorite causes, she gave a glittering reception and dinner in his honor that was designed to introduce him to influential prospects. Andrews recalled that her house "was a blaze of lights and flowers. Twenty people were at the table among whom I remember Thomas Lamont and Dwight Morrow of J. P. Morgan and Company, William Boyce Thompson, the international mine owner, Clarence Mackay, President of the Postal Telegraph Company, and the distinguished, snowy-haired Mrs. E. H. Harriman." After dinner, about thirty more guests assembled in the ballroom where Andrews, using lantern slides from his previous trips to Mongolia, expounded on his plans for the expedition. "Always I stressed the point that it was a gamble. No one knew what was there, if anything, but the scientific dividends [could] be enormous."

As a result of the evening, Andrews received a flood of invitations to dinner at "dozens of New York's greatest houses. Almost every night," he exclaimed, "I donned white tie and tails, climbed into a taxi, and [appeared] at some function where the cards read: 'To Meet Roy Chapman Andrews.'" He once attended thirty-two dinners in succession, and almost every week he spoke at two or three luncheons. Yvette, with her engaging charm, patrician upbringing among European aristocracy, and extensive travels in China and Mongolia with Andrews, was the perfect companion to help her husband navigate the intricacies of these social events.

"Rushing about all day," he complained, "staying up until one or two every morning, keyed to a nervous pitch, almost wore me out at times, but never quite. Sometimes it seemed that I'd give my soul to eat crackers and milk and go to bed just one evening instead of dining on caviar, green turtle soup, and roast duck." Even so, this outpouring of attention was a vindication of Andrews' original strategy: the idea that the expedition could be launched by making it "fashionable" with wealthy supporters was proving to be correct.

Amid his frenetic social schedule, Andrews managed to write his third book, *Across Mongolian Plains*, a lively account of the journey he and Yvette had made in 1919 on the Second Asiatic Expedition. Published in 1921, and illustrated with Yvette's photographs, the volume brimmed with Andrews' fascination for Mongolia, especially what he envisioned as its potential for broadly based scientific research.

With no imminent prospects for large donations in sight, Osborn and Andrews decided to give a "bang-up" men's party at the University Club, one of the most exclusive bastions of the city's rich and famous. After meticulously refining the guest list, printed invitations went out in Osborn's name to fourteen of New York's financial moguls. Among those who assembled in the club's beautifully appointed Council Room were E. H. Gary, president of the United States Steel Corporation; Henry P. Davidson and Thomas Lamont of J. P. Morgan and Company; Sidney M. Colgate, vice president of the soap, perfume, and toothpaste empire; Arthur Curtiss James; Childs Frick, heir to his father's steel and coal fortune and a paleontologist with close ties to the American Museum; John T. Pratt of Standard Oil; George F. Baker, president of the First National Bank; J. P. Morgan, Jr.; and Cleveland H. Dodge, of the Phelps-Dodge Corporation, which controlled enormous holdings in metals, mining, and railroads. John D. Rockefeller, Jr., was unable to attend, but sent his friend and legal adviser Starr J. Murphy in his place. "Probably there was enough money represented in that room," Andrews wrote, "to buy a quarter of the United States."

Using his slides from Mongolia, Andrews lectured for an hour after dinner, then spent another hour answering questions from his engrossed audience. Before the evening ended, George F. Baker, taking Andrews aside, asked him to make an appointment at his office at the First National Bank. "Bring your maps," he added. "I love maps." Several days later, Andrews arranged to meet with Baker, a man noted for his reticence and dislike of publicity (whose personal fortune, derived primarily from banking, was valued in excess of $60 million at the time of his death in 1931). Baker pored over Andrews' map of central Asia, questioning him about every detail of the expedition's logistics and objectives. When he had finished, Baker picked up a telephone and instructed his secretary to "Draw a check in Mr. Andrews' favor for twenty-five thousand dollars."

At this point, Andrews began to sense victory. Donations ranging from $1,000 to $10,000 flooded in from Sidney M. Colgate, Henry P. Davidson, Cleveland H. Dodge, Childs Frick, W. A. Harriman, Thomas Lamont, Clarence MacKay, Arthur E. Newbold, Jr., John T. Pratt, Mrs. Arthur Ryerson, Julius F. Stone, Mr. and Mrs. Charles L. Bernheimer (Andrews' loyal supporters from earlier expeditions), as well as other wealthy New Yorkers who, exactly as Andrews had predicted, now regarded sponsorship of the venture as a social obligation.

Encouraged by his success, Andrews resolved to approach John D. Rockefeller, Jr. Osborn was not optimistic. Having repeatedly failed to induce Rockefeller to visit the Museum, Osborn felt Andrews' chances of interesting him in the expedition were equally slim. But Andrews had already set the stage indirectly through Starr Murphy, the distinguished lawyer and Rockefeller confidant who had attended the gathering at the University Club (for years, Murphy was chief legal counsel for the Rockefeller Foundation). During his lecture, Andrews observed that Murphy appeared keenly interested, and two days later he made an appointment to see him. "He had assigned me fifteen minutes, but I was with him more than an hour. Mr. Murphy was a scholarly man, deeply interested in evolution and the possibility that we might find early human types in the Gobi Desert." He agreed to speak to Rockefeller on Andrews' behalf, and promised to advise him of the outcome.

After ten days, Murphy telephoned Andrews with word that Rockefeller would like to see him. Andrews, who had never met or even seen Rockefeller, was put at ease instantly by his relaxed manner. While Andrews spoke of his plans, Rockefeller listened quietly, showing no outward reaction. "I left without the slightest idea of whether or not I had made a favorable impression," Andrews remembered. Two weeks had passed when a terse note arrived from Rockefeller's secretary: "Mr. John D. Rockefeller, Jr. directs me to inform you that he will be pleased to contribute fifty thousand dollars to the Asiatic expeditions. He wishes to know how the money will be paid."

With Rockefeller's gift, the expedition's funding had reached the $200,000 mark, and Andrews soon prevailed upon other donors to make up the $50,000 shortage. His "society expedition" was about to become a reality—a remarkable testament to Osborn's powerful connections and Andrews' mastery at conveying his irrepressible enthusiasm to others in ways that left them mesmerized. One journalist extolled Andrews' "sheer genius in convincing pragmatic businessmen like Morgan, James, Rockefeller, Baker, and Dodge to underwrite a risky and decidedly nebulous scientific quest in the Gobi Desert." Andrews suggested that whatever else may have contributed to his success, he had managed to capitalize on his suspicion that most financiers were adventurers at heart, whether or not they were consciously aware of it. "Making their money had been an adventure," he declared, "and perhaps they relished a bit of adventure in spending it." If his theory was correct, he was about to satisfy their vicarious yearnings on a grand scale.

Almost on the eve of Andrews' departure for Peking, he was sorely tempted by two offers that would have provided more than enough money to make up the expedition's deficit. The vice president of a major oil company (which Andrews did not identify) offered him $50,000 if he would take along one of its geologists, who would act, he said, "entirely under my orders but still do a spot of looking for oil." A week later, a similar proposition was made by a mining syndicate interested in possible mineral reserves in the Gobi. Andrews rejected both offers. "I was determined," he wrote, "that the expedition should be strictly scientific and its objectives be only what we said they were without taint of commercialism. There could be no 'secret covenants secretly arrived at. . . .'

It never occurred to me that because we had for backers such men as John D. Rockefeller, Jr., J. P. Morgan, Jr., Cleveland H. Dodge, and other names synonymous with vast oil and mineral interests that the expedition would immediately be under suspicion, not only abroad but at home." Accusations that some of its sponsors harbored ulterior motives in supporting the project would continue to circulate for years.

CHAPTER 8

With his financial backing assured, Andrews decided to make a public announcement concerning his plans. It was scheduled to coincide with the appearance of an article, "A New Search for the Oldest Man," which he had written for the November 1920 issue of *Asia*. "We carefully synchronized the press release," Andrews wrote, "with the day the magazine appeared on the newsstands." On the afternoon of October 18—*Asia* was due out the following morning—representatives of twenty-one news organizations gathered in the Members' Room at the Museum to meet with Andrews, Osborn, and the director, Frederick Lucas.

The meeting opened with a briefing on the expedition's attempt to verify Osborn's hypothesis regarding Asia's role in the evolution of mammals and humanity's precursors. Andrews then explained his blueprint for testing this premise through a correlated study of Mongolia's geology, paleontology, archaeology, zoology, botany, and ancient climate, together with a program to map and photograph the region. And he outlined his experimental strategy for using automobiles supported by a supply caravan, as well as the concept of dividing the expedition into self-sufficient units that could roam the desert more or less at will—revelations that triggered a flood of skepticism as to the feasibility of driving automobiles over vast stretches of hazardous terrain without roads or accurate maps.

Almost from the beginning, the press conference was dominated by what Andrews disgustedly called "the curse of the Missing Link." Every newsman present was captivated by the thought of searching the Gobi Desert for traces of mankind's ancestors, aware that such a controversial quest, conducted in one of the most mysterious corners of the earth, was guaranteed to sell papers. Although Andrews emphasized that he was going to Mongolia with the *hope* and not necessarily the *expectation* of finding fossil anthropoids, questions on this subject dragged on interminably and with predictable consequences.

"We hoped the announcement would make the front pages in the morning," Andrews recalled, "and sure enough, it got [a] place of honor in every New York newspaper, and the press wire agencies sent it all over the world." Yet to Andrews' dismay, editors lost no time in sensationalizing his hunt for early humans, or, more precisely, as the papers dubbed them, "ape-men," "the Missing Link," "dawn-men," and other lurid-sounding creatures. Overnight, these fanciful beings became the centerpiece of every news story. A typical example was the *Washington Post*'s headline: WILL SEARCH ASIA FOR "MISSING LINK": EXPEDITION TO SEEK REMAINS OF NEAR-MAN IN GOBI WILDS FOR FIVE YEARS. Even the *New York Times* joined in the fervor when it announced SCIENTISTS TO SEEK APE-MAN'S BONES: NATURAL HISTORY MUSEUM WILL BEGIN FIVE-YEAR QUEST FOR MISSING LINK. . . .

At first, Andrews was furious, afraid that so much emphasis on "ape-men" might detract from the expedition's other objectives and discredit him among scientists already dubious about his chances for success. "In vain," he wrote, "did I try to direct attention to the larger aspects of our work: the test of Professor Osborn's theory of Central Asia as a theater of mammalian evolution." Osborn's pronouncements concerning mankind's Asiatic genesis were hypothetical, he reiterated, and uncovering hard scientific evidence such as skeletons and implements to prove that ancient hominids once inhabited the region "was much like looking for the proverbial needle in a haystack. . . . All we could do was hope." But this sober reality "didn't create a ripple in the newspaper world," Andrews complained. "Primitive man was what they wanted and anything else bored them exceedingly. In a week, we were known as the 'Missing Link Expedition.'"

The phrase caught the public's imagination "like fire in dry leaves," Andrews wrote. "I was flooded with [crank] letters. . . . 'Why go to Asia to hunt the Missing Link,' someone inquired, 'I saw him in the subway this morning.' That was the standard joke, but there were dozens of others. One woman telegraphed: 'Regarding search for Missing Link, Ouija Board offers assistance.'" The outpouring of publicity surrounding the project became so overwhelming that Andrews eventually resigned himself to its sensational aspects. Evolution was an emotionally charged subject—then as it is today. With names like Morgan, Rockefeller, Baker, Frick, and James among the expedition's backers, added to the fact that it was organized under the auspices of a respected scientific institution and led by an explorer already famous for his adventurous exploits, the determination of the press to capitalize on this all-out hunt for ancient man was inevitable.

Racism was another factor in the rush to pinpoint Asia as man's birthplace, adding to the intense controversy that engulfed Andrews' venture. At that point, the earliest skeletal remains that could be classified as anthropoid came from a site along the Solo River near Trinil in Java. Here, in 1891, a Dutch physician, Eugène Dubois, uncovered a skull fragment, several teeth, and a femur belonging to a creature he named *Pithecanthropus erectus*. Its skullcap and teeth were definitely apelike, but the femur, the upper leg or thighbone, was indistinguishable from that of modern man, indicating that *Pithecanthropus* had walked upright with a two-legged gait. What's more, the age of this so-called "Java Man" was initially dated by geological strata at about 600,000 years, much older than any of the Neanderthal or Cro-Magnon skeletons found in Europe.

With Dubois' discovery, Asia—rather than Africa, as some scientists had suggested—assumed center stage as the place where humans began their climb up the evolutionary ladder. To those anthropologists for whom questions of evolution and race were inseparable, this was welcome news. Like many of his colleagues, Osborn espoused the superiority of Caucasian peoples, especially those of Nordic stock. The possibility that man's genesis might have occurred in Africa, the "Dark Continent," with its Negroid stigma, was decidedly unpalatable. Asia, therefore, was far preferable as the "Cradle of Mankind," and to Osborn and Andrews proving this hypothesis was almost a sacred duty.

Osborn's views on race, which Andrews seems to have embraced uncritically, reflected a variety of concerns evidenced in the United States by a vigorous campaign for eugenics. Its adherents sought to combat what they regarded as a threat to the nation's identity and stability by advocating laws aimed at sterilizing persons with genetic mental or physical impairments, restricting the immigration of "undesirable" ethnic groups, preventing interracial marriage, and encouraging higher birth rates among the most intelligent and productive members of society.

Throughout the early 1900s, until World War II, in fact, the eugenics movement was supported by legions of scientists, teachers, doctors, and politicians. (Two early advocates were Mrs. E. H. Harriman and Franklin D. Roosevelt.) Osborn joined with several of his trustees at the Museum in organizing the Galton Society, which worked to foster the principles of eugenics. During the early 1930s, Osborn went so far as to endorse Hitler's philosophy of racial purity, and in 1934, he even traveled to Germany to accept an honorary doctorate from Johann Wolfgang von Goethe University for his contribution to eugenics.

Nor was it surprising that the mere mention of the Missing Link or ape-men, indeed, the issue of human evolution in general, subjected the expedition to the wrath of Christian Fundamentalists, adding even more to its newsworthiness. The suggestion that mankind "evolved" from lower anthropoids, whether in Asia or anywhere else, raised the hackles of reactionaries within many religious groups who viewed Andrews' project as a well-orchestrated plot to promote Darwin's concept of evolution as opposed to the biblical version of creation. Initiated by a hard core of hellfire-and-brimstone preachers, overwhelmingly Protestant and galvanized by the fiery orator and thrice-defeated Democratic presidential candidate, William Jennings Bryan, the antievolution movement had surged to the forefront of America's political arena shortly before World War I.

Almost from the day the Museum announced plans to search for primitive man in Mongolia, Osborn was savagely vilified by Fundamentalists because of his presidency of the Museum and enormous influence as an author and educator on the subject of evolution. So fierce was the onslaught against Osborn by his arch-adversary Bryan—who four years later would prosecute John Thomas Scopes for teaching evolution in a Tennessee school—that beginning in 1922, Osborn engaged Bryan in a series of heated debates, widely publicized in newspapers and magazines. His retorts to Bryan, some of which appeared in the *New York Times*, were later collected in a book titled *The Earth Speaks to Bryan* (1925) and distributed to scientists enlisted by John Scopes' defense attorney, Clarence Darrow.

No sooner had word of the expedition hit the newspapers before droves of people applied to go along. Andrews was inundated with so many letters, telephone calls, and cables that printed cards had to be mailed to aspiring applicants informing them that all staff positions had been filled but their requests would be kept on file. Numerous inquiries came from former soldiers looking for adventure, professional men seeking escape from the boredom of their daily routines, and teenage boys to whom Andrews had become a hero of almost fictional dimensions. One man offered his services as a "barber and hairdresser (also sharp-shooter)"; another, a butcher by trade, sought a position as Andrews' bodyguard. A waiter in a New York restaurant wanted "to serve your tables in the Gobi," adding as an extra inducement, "I own a tuxedo. So you will not have to furnish that."

Many letters were from psychics and fortune-tellers. A woman in Missouri assured Andrews that spirits had revealed the whereabouts of a city buried under the Gobi's sands that contained a record of man's development "from the time he crawled on all fours to the dawn of history." Amused, Andrews wrote back to ask if the spirits could be more specific regarding the city's location, since the Gobi covered a very large area. Two weeks later, he received a reply saying that his inquiry had annoyed the spirits, who would only add that the city was marked with four stones half-buried in the sand.

Out of approximately ten thousand applications received during the expedition's first five years, three thousand were from women, some of whom resorted to sexual innuendos hoping to bolster their chances. One particularly amusing letter read in part:

Dear Doctor Andrews:

I want to apply for a position as secretary on your next expedition. . . .

I am looking for something occult and stirring and I think I can find it with you. I have seen your picture in the newspapers and I know from the kindness and nobility of your face that you know how to treat a lady.

If no position of secretary is open, perhaps you can take me just as a woman friend. I could create [a] home atmosphere for

you in those drear wastes. I am sending my photograph, but it is
much better to see the original. How would Friday afternoon do
for tea?

A note from a woman in New York City railed against "laws passed
by men" who sought to define women's proper place in society. Vowing
never to accept such restrictions, she announced her intention of blazing her own trails:

> . . . I want to do something worthwhile . . . something that
> will make me feel worthy of life. . . . Now what I want to ask is:
> Can a lady organize an expedition or is it possible for her to join
> one? If you should answer please don't try to tell me of the dan-
> gers due to sexual feelings because they do not exist in me and
> cannot be aroused so that ends that argument.

Even as Andrews' fund-raising drive was in progress, he had already begun selecting his scientific staff and assembling equipment. Next to fi-
nances, choosing the men who would accompany him to Mongolia
was his highest priority. Apart from possessing the requisite professional
expertise, it was also essential that every member of his staff was physically fit, able to cope with unexpected hardships, and compatible with
everyone else.

For his second-in-command and chief paleontologist, Andrews enlisted his longtime friend at the Museum, Walter Granger, who was
then associate curator of fossil mammals. It was an easy choice. Granger
commanded tremendous respect as a scholar and technician. He was a
large, self-effacing New Englander, with a mustache, a hearty laugh,
and a droll sense of humor. Granger's integrity, sound judgment, and
profound knowledge of paleontology made him ideally suited to over-
see the scientific direction of the expedition, leaving Andrews free to
deal with its organizational, diplomatic, logistical, and financial aspects, plus his duties as chief zoologist.

Moreover, the two men complemented each other perfectly. Andrews' flamboyance and headstrong determination were offset by
Granger's objectivity and restraint. Granger possessed the ability to

comprehend Andrews' aggressive, devil-may-care style, even if he did not always condone it. Writing in *Natural History*, D. R. Barton aptly observed that "Andrews was the impetuous type, the driver, his eye turned toward the horizon. Granger, on the other hand, was the conservative—careful and deliberate, with a hound's capacity to nose out a buried bone. Andrews has summed it up with characteristic vigor: 'I was the accelerator,' he says. 'Walter was the brake.'" As Barton also pointed out, Andrews considered Granger "the best natured man he has ever known."

In some ways, their backgrounds at the Museum were similar. Born in Middletown, Vermont, in 1872, Granger developed a passion for natural history while studying wildlife in the woods and hills near his home. With only two years of high school to his credit, he went to New York to work at the Museum. A brief apprenticeship in the taxidermy studio led to an interest in zoology, comparative anatomy, dinosaurs, and fossil mammals. Encouraged by several of the Museum's most respected scholars—Jacob Wortman, Frank M. Chapman (Andrews' boyhood idol), and the redoubtable Henry Fairfield Osborn—Granger acquired a prodigious command of paleontology entirely through laboratory research and field experience rather than formal education. Renowned as a collector of fossils, he participated in numerous expeditions; and, in 1932, Middlebury College in Vermont awarded Granger his only degree, an honorary doctorate of science.

Andrews was equally fortunate in his choice of a chief geologist. He engaged the genial, brilliant Charles Peter Berkey, a professor at Columbia University and a towering figure in the study of petrology, structural geology, paleogeography, and stratigraphy. Berkey's ingenuity in solving geological problems connected with massive construction projects resulted in an extraordinary career. Among other undertakings, he was a consultant for the massive Catskill Aqueduct, the George Washington Bridge across the Hudson River, the Triborough Bridge, Lincoln Tunnel, Saint Lawrence Seaway, Queens-Midtown and Brooklyn-Battery tunnels, Hoover Dam, Grand Coulee Dam, the Panama Canal, and the United Nations Building.

On Berkey's recommendation, Andrews hired another Columbia faculty member, Frederick K. Morris. A structural geologist and former consultant in geography to the Department of State, Morris had the advantage of having taught in China at Pei Yang University in Tientsin, where he learned Chinese. Because Berkey and Morris had been closely associated at Columbia, Andrews was confident the two men would make an effective team in the Gobi, a conviction that proved amply justified.

Andrews planned to limit his first season's activity to reconnoitering the countryside and pinpointing promising areas for future investigation. With the selection of Granger, Berkey, and Morris, he now had the scientific talent to carry out such a preliminary survey, but the expedition still lacked two key technicians without which it could not proceed: a photographer and a mechanic to maintain the automobiles.

Because the American Asiatic Association's sponsorship agreement included exclusive rights to a motion picture on the explorers' adventures, it had a vested interest in ensuring that the footage was exciting and commercially viable; the object was to show it in theaters around the country at a handsome profit. Accordingly, *Asia* magazine's editor, Louis Froelich, was eager to secure the services of the famed cinematographers Martin Johnson and his wife, Osa, who had already earned rave reviews for travel films shot in the South Seas and Africa. At the time Froelich approached the Johnsons about Mongolia, they were committed to a project in East Africa with Carl Akeley. Nevertheless, they tentatively agreed to join Andrews' venture if Froelich could meet their hefty financial demands and they completed filming in Africa in time to reach Peking before the expedition's departure. Despite months of negotiating, no contract was signed and the Johnsons sailed for Africa, promising to resolve the matter by cable from Nairobi.

Before going to Peking in 1921 to prepare for the expedition, Andrews had met another photographer named J. B. Shackelford. They were introduced by George H. Sherwood, the Museum's executive secretary and a member of the Akeley Camera Company's board of directors. Andrews took an instant liking to the jovial, portly Shackelford, who was then working for the Akeley Camera Company in New York. Originally designed to facilitate wildlife photography, Akeley's camera, an ingenious improvement on its competitors, had precision gears and flywheels mounted in the head that allowed it to be turned in any direction, smoothly and with great speed. The lens could be focused rapidly while the camera was running. Equipped with an easily changeable two-hundred-foot film magazine, it was lightweight (forty pounds), could be operated without a tripod, and was practically indestructible. Apart from their ability to capture swiftly moving animals, Akeley's cameras became the first choice of combat photographers during World War I, and for some years afterward, they were widely used by Hollywood filmmakers.

Andrews believed all along that Martin Johnson, though a gifted photographer, was "too unreliable," as he told Froelich, ". . . [and] accustomed to doing things on such a large scale that he would not be satisfied or able to do them on the money which we have available." When Johnson failed to answer cables sent to Nairobi regarding a last-minute financial offer made by *Asia*, an impatient Andrews wired Froelich: "Please drop Johnson, send Shackelford." Bitterly disappointed over the Johnson debacle, Froelich immediately engaged Shackelford. It proved to be a wise decision. As an expert in handling the Akeley camera, Shackelford produced magnificent footage of the expedition, a total of fifty thousand feet, along with hundreds of photographs; he was also amiable and eager for adventure, and he unexpectedly turned into a sharp-eyed fossil hunter.

When it came to outfitting the expedition, Andrews displayed a genius for organization and advance planning that had been finely honed by his previous explorations in Korea, China, and Mongolia. Absolutely nothing was left to chance. He was determined that his staff should be as comfortable as possible in the field, well provisioned, and properly nourished if they were to operate at maximum efficiency.

Andrews was fond of asserting, "Adventures are a mark of incompetence," a dictum coined by the noted Arctic traveler Vilhjalmur Stefansson. "If the explorer has a clear-cut problem to solve," wrote Andrews, "and an honest desire to contribute something of worth . . . he will prepare against adventures. It will disappoint the newspapers but facilitate his work. How infinitely more credible it is to eliminate difficulties through foresight and preparation before they are encountered than to suffer heroically and leave the work half done." Ironically, Stefansson's expeditions were often plagued by misadventures, including a near mutiny by his companions, and his pronouncement linking adventure with incompetence drew angry criticism from some explorers, who felt it was untrue and trivialized unavoidable dangers inherent in venturing into unknown places.

Nonetheless, Andrews set about preparing for his venture convinced that most disasters could be avoided with proper planning, though he would have cause to question this premise once he reached Mongolia. Assisted by manufacturers' representatives and the world's best safari outfitters, Andrews acquired the finest equipment available. His checklist covered twenty-two typewritten pages and included everything one could possibly want during six-month stretches in the desert, even hot water bottles and a phonograph and records. In addition, he took tons of canned and dehydrated vegetables and fruit, flour, sugar, coffee, tea, powdered milk and eggs, canned hams and bacon, lemon powder, cigarettes, and a supply of whiskey and brandy "for special occasions."

Observing that "natives in every country have developed the best type of dwellings and clothing for their particular conditions," Andrews had sheepskin coats, trousers, and sleeping bags made to combat the Gobi's sudden cold spells. He also ordered exact copies of the typical Mongolian tents, made from two layers of blue cotton cloth with sides that sloped down from a center ridgepole and helped deflect the incessant desert winds. The traditional yurts, or *gers*—portable structures of thick felt covered with canvas and supported by a framework of poles— were too large and heavy to be transported by the expedition.

More than any single item of equipment, the success or failure of Andrews' plan depended upon the sturdiness of his automobiles. From his previous experience in driving from Kalgan to Urga, with frequent side trips off the trail in search of game, he wanted a car that was lightweight, with a flexible chassis, and at least a twenty-eight horsepower engine. Eventually, he decided on Dodge Brothers touring cars, which, he said, "filled these specifications to the letter." In addition to three Dodges, he purchased two one-ton Fulton trucks, manufactured by the Fulton Motors Company in Farmingdale, Long Island. While these trucks performed well enough, they proved too heavy and were replaced with Dodges after the 1925 season. With five vehicles to supply, a large part of the caravan's inventory consisted of extra oil, grease, and enormous quantities of gasoline, together with every imaginable spare part and the necessary tools for repairs. "Our motor experts were highly trained men," Andrews noted, "thus short of actually wrecking the chassis or engine, we were prepared for any emergencies."

As chief of motor transport for the first season, Andrews took along S. Bayard Colgate, a scion of the Colgate-Palmolive family. Bayard's father, Sidney M. Colgate, and two of his uncles, Austin and Russell Colgate, were among Andrews' most enthusiastic supporters. A graduate of Yale, Bayard had become interested in automobiles as a youth and was an accomplished mechanic, a skill he further refined by undergoing special training at the Dodge and Fulton factories.

By March of 1921, Andrews' great venture—a small fortune in its coffers, superbly staffed and equipped, and labeled a *de luxe* expedition by the press—had become a reality. Before the explorers could finally set out on their long-anticipated quest, however, Andrews faced another year of preparations in Peking and Urga. Among a long list of unfinished tasks, it remained to establish a headquarters for the expedition; secure permits from the Chinese and Mongolian governments to operate in their territory; select and train native assistants to work as interpreters, cooks, technicians, chauffeurs, and camel drivers, construct crates for transporting provisions and specimens, buy camels for the supply caravan, and purchase additional food. Except for special canned and dehydrated foods brought from America, all other staples were obtained through the American legation in Peking from the U.S. Marine commissary.

As Andrews rushed around New York attending to last-minute details, his elation was tempered, if only momentarily, by criticism from die-hard skeptics, a small but vociferous clique of scientists who predicted that the expedition was headed for catastrophe. It was one thing, they ominously declared, to drive a well-traveled road from Kalgan to Urga, but quite another to take automobiles across the Gobi's treacherous expanses. And the concept of relying for resupply on a slow-moving caravan, vulnerable to bandit attacks, was deemed too risky, raising the possibility that the explorers would end up stranded. Andrews was keenly aware of these dangers, but there was no other way to get supplies into the field and transport specimens back to Peking. The obvious solution—the use of airplanes, which Andrews had considered—was forbidden by Chinese officials, who, following World War I, associated aircraft with military activities and espionage.

Even without these obstacles, the expedition would have to confront the Gobi's unpredictable climate, lack of reliable maps, and rampant political unrest. Yet what most troubled Andrews were critics who insisted that Mongolia was devoid of fossils and was essentially a scientific wasteland. "When it was announced that we were to explore the Gobi Desert for fossils," he wrote, ". . . our project was ridiculed by some . . . scientists, particularly in Peking. They pointed out that we would find only a waste of sand and gravel and that we might as well search for fossils in the Pacific Ocean. . . . Also I was told it was little short of criminal to squander the time of such eminent geologists as Berkey and Morris in a country where the geology 'was all obscured by sand.'"

Writing in his 1929 memoir, *Ends of the Earth*, Andrews confessed that even he suffered twinges of anxiety over what lay ahead. "No one realized more fully than I what a chance we took. If the expedition were a failure I never could face those loyal men and women I had persuaded to back it with their faith and dollars. In that event I knew it would be my 'swan song' in exploration. I remarked as much to Professor Osborn at our last meeting." Osborn's response was characteristically dramatic. "He put his hand on my shoulder," Andrews recalled, and said, "'Nonsense, Roy. The fossils are there. I know they are. Go and find them.'"

In February 1921, Andrews, Yvette, and their three-year-old son, George, left New York for San Francisco. From there, they sailed to China aboard a new ship, the *Golden State*, taking with them thirty-eight tons of equipment and supplies. After a year of unrelenting fund-raising activities, interviews, lectures, and planning, Andrews was exhausted and "almost a nervous wreck. . . . [But] after all it was worth the price," he wrote. "The expedition of my dreams was an accomplished fact." It ventured forth bearing a designation worthy of its size and scope: THE THIRD ASIATIC EXPEDITION IN COOPERATION WITH THE AMERICAN ASIATIC ASSOCIATION AND *ASIA* MAGAZINE. Andrews' odysseys into Yunnan in 1916–1917 and Mongolia in 1919 had constituted the First and Second Asiatic Expeditions respectively. Later, in 1925, the project was renamed THE CENTRAL ASIATIC EXPEDITIONS.

CHAPTER 9

Not since 1900 had Peking experienced a dust storm equal to the one that greeted Andrews and Yvette when they arrived on April 14. Whipped by winds sweeping across the arid plateau of Mongolia and northwestern China after fourteen rainless months, its yellow haze reached as far south as Shanghai and hovered sixty-five miles out to sea. One could scarcely distinguish Peking's ancient Tartar walls as Andrews' train from Tientsin pulled into the station. By now, the Chinese had come to regard the curtain of dust that hung over the city as an evil omen, boding a summer of famine, disease, and war. Adding to the atmosphere of tension, a smallpox epidemic had broken out in the interior of China, and Peking was buzzing with rumors of an impending clash between two of the country's most powerful warlords, who were vying for control of northern China.

"The same dear old hysterical Peking!" Andrews commented gleefully. "[We foreigners] were a rather small community and excitement was *sine qua non*. If no political bomb was ready for explosion, something must be manufactured to furnish conversation at the Club or on the roof garden of the new hotel. So with dust, war, and smallpox, we felt the summer was beginning rather well."

Inherent in Andrews' seemingly callous remarks was the fact that everyday realities of life in China, as harsh as they were for the masses, seldom touched the members of Peking's foreign colony. Insulated by their legations, renovated Manchu palaces, charming bungalows, clubs, and servants, the roughly two thousand Europeans and Americans who lived there permanently looked with detachment upon the wretched plight of most Chinese. Rarely were they concerned with the political quagmire that surrounded them, which frequently assumed the qualities of a comic opera.

Even when convulsed by turmoil, Peking's seductive charms endured: its magnificent gardens and flower markets; its treasure-laden shops; the omnipresent fragrance of incense; the wandering troups of jugglers, acrobats, and magicians; flocks of pigeons with tiny whistles attached to their tails; the ghostly splendor of the Forbidden City, virtually deserted except for the quarters occupied by the pathetic last emperor, Pu Yi, and his retinue; crowded streets where coolies and rickshaw boys vied with loaded camels and automobiles; a cryptic aura of mysticism that permeated the narrow streets and alleyways, all of which conspired to make Peking what one writer called "the *femme fatale* of cities."

Like the denizens of all expatriate enclaves, their social life centered around a never-ending whirl of diversions—in this case, tennis matches, badminton, horse racing, steeplechases, polo, hunting, cocktail parties, dinners, receptions, bridge and mah-jongg tournaments, all-night dances, champagne breakfasts, and picnics in the Western Hills, along with a constant stream of gossip and sexual intrigue.

Haunted perhaps by the realization of how rapidly revolutionary sentiments among the Chinese were threatening the privileges they had enjoyed under the Manchus, foreigners pursued their frivolities with manic compulsiveness. Their attitude was perfectly exemplified by an episode that occurred in 1922 when a detachment of Wu Pei-fu's army camped on a racecourse used by the Peking Hunt Club. Enraged by this inconvenience ("Anger mounted with every cocktail," Andrews quipped), the club sent a delegation of stewards to present Wu Pei-fu with a formal protest. Although the general agreed to move his troops off the track, he relocated them a short distance away where they still crowded the road leading to the racecourse. When riders appeared early the next morning to exercise their horses, they were greeted by a soldier carrying a sign printed in english that read, PLEASE MAKE DETOUR. THIS WAY HAVE GOT ONE WAR. The reaction of the club's members to the incident was one of fierce indignation. "If the Chinese wanted to have a war," Andrews railed, "and it amused them to do so, all right. But it was quite another thing when they let their dashed war interfere with our race meet. It was a bit too thick and we wouldn't stand for it!"

However much outsiders tended to dismiss China's internal up-heavals as opéra bouffe, the country's anguish reflected a protracted struggle to throw off centuries-old feudal traditions, free itself from imperialistic domination, and find a new national identity. The lamentable state of affairs that awaited Andrews in 1922 was an outgrowth of the chaos that followed Yuan Shi-kai's ill-fated attempt to install himself as emperor in 1915. Although Yuan held the throne barely a hundred days and died shortly afterward, it was too late to restore order or reverse the growing animosity toward the central government in Peking. One province after another fell victim to the "strongman" ideology embraced by ambitious military officers, many of them financed by the opium trade, who used the breakdown of the old regime to elevate themselves to the status of warlords, grabbing political power, territory, and spoils in the process. For those among this clique with sufficiently large egos and powerful-enough armies, the control of China itself became their objective, and the consequences of such adventurism were disastrous. Assassinations, banditry, looting, ever-changing alliances, betrayal, forced conscription of soldiers, bribery, anarchy, and perpetual warfare were rampant.

By the time Andrews reached Peking to begin preparing for his expedition, the two strongest warlords in northern China were Chang Tso-lin and Wu Pei-fu. The pompous, arrogant Chang Tso-lin was a former Manchurian bandit, who had switched to a military career, amassed a fortune during the Russo-Japanese War by selling supplies to both sides, and eventually gained control of Manchuria, with a crack army under his command. By contrast, Wu Pei-fu, diminutive, outwardly shy, and fond of writing poetry and painting, came from an educated family in Shantung Province. After studying with private tutors, he earned a bachelor of arts degree and entered military school, where he displayed a remarkable talent for strategy and leadership. And looming in the background was a third figure who was destined to take his place among China's major warlords—a hulking, puritanical officer named Feng Yu-hsiang, a fanatical Christian convert who would desert in 1924 as one of Wu Pei-fu's trusted commanders, join with Chang Tso-lin to defeat Wu, and thereby begin a power play of his own.

Andrews' first priority was to establish a headquarters for the expedition, a task facilitated by the fact that he had already selected an ideal residence for this purpose. It was previously occupied by Andrews' late friend George Morrison, a correspondent for the *London Times*. Formerly a Manchu palace, the house was located at 2 Kung Hsien Hutung (Bowstring Street), northeast of the Forbidden City in a district known as the Tartar City. It was completely hidden behind a high wall, with enormous, brass-studded gates that opened from the *hutung*—the narrow, alleylike streets that crisscross Peking's residential areas—into a lovely courtyard sprinkled with trees. Sprawled over an acre of land, the original palace compound consisted of 161 rooms built around ten courtyards. Until the demise of the Manchus in 1911, a prince had lived there in regal splendor surrounded by servants, retainers, and at least thirty concubines. When Morrison rented the palace, he pulled down two buildings and tore out a number of walls, reducing it to forty-seven rooms and eight courtyards, but after his death in 1920, the property had fallen into a state of disrepair.

Andrews engaged an intermediary whose job was to arrange details of the lease between the landlord and tenant. In theory, he was also supposed to handle the never-ending demands for "squeeze" or bribes, without which nothing could be accomplished. After five frustrating weeks of negotiating, Andrews succeeded in leasing the property, but he was left to confront the intricacies of squeeze on his own. In addition to regular charges for their services, everyone Andrews dealt with— including the intermediary—extracted the mandatory squeeze: the equivalent of a month's rent went to the police for a permit to move in on schedule, $45 extra to the water company, and $100 to the electric company "for the privilege of using their light."

"Thus it went through every phase of getting the house ready to occupy," Andrews wrote. "Squeeze, squeeze, squeeze! You rebel and say that you won't pay. All right, you live in a hotel. You might as well realize at the beginning that you will pay squeeze on every item that comes into your house from five coppers worth of carrots to the rugs on your floor. Somebody gets his commission whether you like it or not. You can't live in China without paying squeeze. It is a national custom."

Once the lease had been signed, an army of carpenters, masons, electricians, and plumbers went to work converting the house to forty rooms that included servants' quarters, three bathrooms, six garages, and stables, plus special features needed for the expedition—a laboratory, an office, a photographic darkroom, and storage areas. At the rear of the compound, enclosing a beautifully landscaped inner courtyard, was a sumptuous residential suite furnished with comfortable sofas and chairs upholstered in silk, ornately carved tables and cabinets, heavy damask curtains, and a superb collection of Oriental *objets d'art*. While serving as the center of activities for the expeditions, the compound would also be Andrews' home for the next twelve years. It remained a well-known landmark until 1982, when it was demolished to make space for an apartment building.

Altogether, Andrews' household staff usually numbered between eighteen and twenty servants. At the top of this hierarchy was a man named T. V. Lo, who held the coveted position of Number One Boy. Intelligent and utterly devoted to Andrews and Yvette, Lo was responsible for supervising the cooks, housekeepers, grooms, gardeners, and gatemen. He also policed the elaborate system of squeeze worked out between themselves. Wages for servants varied somewhat according to the size of individual houses, but Lo received what was considered an excellent salary—$10 a month (gold); the rest of the indoor servants were paid about $7; and the coolies who worked outside got $5.

Along with outfitting the headquarters, plunging into Peking's feverish social life, and playing polo, Andrews spent several hours a day studying at the North Chinese Language School, improving his command of the Mandarin tongue—the country's official lingua franca. He also began selecting and training Chinese assistants who would accompany the expedition as chauffeurs, cooks, and collectors. And he set about the formidable task of unpacking and organizing the thirty-eight tons of equipment, which included the three Dodge touring cars and two Fulton trucks, brought from the United States, while purchasing additional provisions through the Marine commissary in Peking. "I never realized how much food twenty-six men could eat in five months," he observed, "until I began to assemble the supplies. A ton and a half of flour, six hundred pounds of sugar, half a ton of rice, one hundred pounds of coffee, two hundred tins of jam and other staples [all of which] made a mountainous heap on the laboratory floor."

Another priority was the securing of permits that allowed the expedition's scientists to collect zoological and paleontological material inside China during the winter, when they were unable to work in Mongolia because of the harsh climate. Upon reaching Peking, Andrews initiated contact with the National Geological Survey, a government-supported organization directed by a respected geologist named V. K. Ting, with the assistance of an American paleontologist, Amadeus W. Grabau, and a Swedish geologist and mining expert, J. G. Andersson.

The Survey's main functions were to conduct geological and paleontological research, locate mineral resources, and train students in these fields. Andrews knew that the Survey's endorsement was crucial to the success of his plans, and he structured an agreement with Ting and his colleagues to guarantee that his expedition did not infringe on locations in China already marked for investigation by the Geological Survey. As a result of the cordial relations generated by this cooperative approach, an area near Wanhsien in eastern Szechwan Province—one known to be rich in fossils—was placed under Andrews' jurisdiction. In turn, Amadeus Grabau, formerly a professor of paleontology at Columbia University, was invited to represent the Survey as a research associate with the Third Asiatic Expedition.

Gaining access to Mongolia was to prove far more difficult. Beginning early in 1920, the government in Urga had been torn apart by a series of bizarre political events that presaged a Communist takeover. Most of Andrews' contacts within the old ruling hierarchy, whom he had counted on to clear the way for his explorations, had been deposed in the heat of Mongolia's revolutionary fervor. Andrews might never have been able to proceed had it not been for his friend F. A. Larsen— the genial Swede who had given Andrews his beloved horse, Kublai Khan, during his 1919 zoological expedition to Mongolia.

Despite Mongolia's recent civil strife, Larsen, who still commanded enormous respect in Urga, had established cordial relations with several newly installed government ministers. Andrews, therefore, asked Larsen to act as his personal representative in obtaining permission for the expedition to work in Mongolia. But after consulting with various officials, Larsen notified him that passports could not be issued until he reached Urga in the spring of 1922, at which time an agreement giving the explorers permission to operate within the country's borders would be drawn up. This meant the expedition would have to enter Mongolia without a firm commitment from authorities allowing it to travel beyond Urga, a potentially disastrous situation that was understandably not to Andrews' liking.

On June 28, 1921, Andrews' chief paleontologist and second-in-command, Walter Granger, arrived in Peking from New York. With the start of explorations in Mongolia still almost ten months away, Granger, after consulting with V. K. Ting and J. G. Andersson of the Geological Survey, decided to investigate the fossil deposits around Wanhsien in Szechwan Province, which Ting had previously turned over to Andrews for exploration. Unfortunately, to reach Wanhsien it was necessary to travel up the Yangtze River into an area embroiled in warfare between the forces of Wu Pei-fu and Szechwanese militia, a fact that would put Granger and his party at considerable risk.

Ignoring this threat, Granger left Peking on August 24. Because Granger did not speak Chinese and was unfamiliar with the country, Andrews sent along an interpreter and field assistant named James Wong, a congenial young man who had attended a military academy in Massachusetts, spoke fluent English, and proved invaluable in dealing with endless problems. In addition, the group included a Number One Boy known as Chow; Yang, an excellent cook; Chih, a taxidermist; and Liu and Van, who acted as all-around assistants.

After reaching Hankow by rail, Granger and his party boarded a ship that carried them through the Yangtze Gorge to Wanhsien. Steaming upriver, they were repeatedly fired upon by soldiers deployed along the banks or crowded onto junks that patrolled the Yangtze. No one was injured, although at times gunfire sent everyone scurrying for cover, and once a bullet narrowly missed Granger as he sat on the afterdeck watching the fighting on shore.

The focus of Granger's investigations was an area near the village of Yen-ching-kou, about twenty miles from Wanhsien. Here, large numbers of fossil mammals turned up in deep natural pits along the top of a limestone ridge into which the animals had apparently fallen during the early Pleistocene epoch, around 1.5 million years ago. Local inhabitants regularly "mined" these deposits, selling the fossils, or "dragon bones" (called *lung ku*), to wholesalers who distributed them to apothecary shops in distant cities. As the Chinese believed that such bones were imbued with supernatural properties, they ground them into a powder for use as medicine, creating a huge market for fossils. Actually, most of what was then known about the paleontology of China had been pieced together by studying fossils purchased from druggists; and a German scientist, Max Schlosser, had drawn upon a large private collection acquired in pharmacies throughout China to make a comprehensive survey of the region's extinct mammals.

Aware that it was impossible to excavate the pits himself, Granger hit upon a simpler method of collecting. After locating all the producing pits on the ridge, he and his assistants made frequent visits to each one, watching carefully as the diggers used ropes and pulleys to descend into the openings and retrieve fossils, which were hauled up in baskets. Granger would then examine the bounty and purchase whatever bones he wanted, at a fixed rate per *gin* (one and a third pounds). Using this method, he soon acquired hundreds of specimens representing a broad spectrum of Pleistocene fauna that included *Stegodon* (a distant relative of elephants), rhinoceroses, bears, tapirs, rodents, gaurs (a large bovid similar to oxen), dogs, and a species of cat "hardly distinguishable," Granger observed, "from the tiger." By the end of his stay at Yen-ching-kou in February 1922, the pits had yielded such a profusion of fossils that Granger returned during the winters of 1922–1923 and 1925–1926 with equally rewarding results.

While Granger was away in Szechwan, Andrews made a short excursion to a mountainous, thickly wooded hunting preserve near the Ming Tombs, about eighty miles northeast of Peking. With him went two Chinese assistants who were being trained to collect and prepare zoological specimens. He was also accompanied by the newest addition to his scientific staff, a herpetologist named Clifford Pope who had arrived in Peking with Granger on June 28. A native of Georgia, the twenty-two-year-old Pope had joined the expedition after graduating from the University of Virginia and working for two seasons in British Guiana under the famous naturalist William Beebe of the New York Zoological Society.

Andrews used the trip to the Ming Tombs to instruct Pope in an expedient method of collecting whereby local inhabitants were recruited to find specimens, an idea Andrews hit upon during his previous travels in China. Scarcely had he and Pope set up camp before crowds of curious onlookers from nearby villages appeared to greet them. After observing the proper courtesies and explaining the purpose of their visit, Andrews offered to pay 3 coppers (about 1 cent) for every frog, lizard, or fish they caught, and a slightly higher price for snakes. At first, the thought of buying animals not intended to be eaten met with total disbelief, but two young boys finally vanished into the woods and returned a short time later with several frogs. When Andrews gave them the agreed amount, everyone rushed off to scour the area, turning up dozens of specimens within a few hours. Once Andrews and Pope had acquired a sufficient number of the more plentiful species, they stopped buying them and offered a premium for the less-common types. Despite drenching rainstorms that hampered the search, the villagers eventually brought in over four thousand specimens. "We left with a confident feeling that we had a complete representation of the [reptilian] fauna," Andrews wrote. "What those two or three hundred Chinese did not find for us must have been very rare indeed!"

Regardless of Pope's youth and relative lack of experience, Andrews never doubted his abilities. Not only did Pope justify Andrews' expectations, he went on to become one of the world's leading herpetologists, holding curatorial positions at the American Museum and the Field Museum. Eventually, he learned to speak fluent Chinese and gained a profound familiarity with the country and its people. Although he never went to Mongolia, due to the scarcity of herpetological specimens in the desert, he spent the years from 1921 through 1923 collecting in the provinces of Anhwei, Hunan, Shansi, and on the island of Hainan. For two years beginning in 1925, Pope returned to China, working mainly in Kiangsi and Fukien. During the thousands of miles he traveled about the countryside, often in extremely remote areas, he withstood hardships, illness, bandit attacks, and civil wars. Accompanied by an entourage of servants, taxidermists, and a gifted artist, Wong Hao-tin, who painted masterful watercolors of specimens that are still among the American Museum's prized possessions, Pope's journeys netted the most comprehensive collection of reptiles, amphibians, and fish ever brought out of China up to that time—over six thousand specimens.

Soon after returning from the Ming Tombs, Andrews set off on an arduous quest for one of the rarest Asian mammals, the takin, or "golden fleece," which the Chinese called *yeh niu* (wild cow). These odd-looking creatures, described by Andrews as "beasts out of the pages of mythology with Roman noses, short horns, and long yellow-gold fur," are closely related to the musk ox, and inhabited the Tsinling Mountains of Shensi Province as well as the highlands of India and Tibet. Eager to acquire specimens of these extraordinary animals for the Museum, Andrews and an English companion, Captain W. F. Collins, an experienced hunter of takin, left for Shensi on September 8, 1921, accompanied by the usual complement of servants and trackers. Ahead lay an exhausting journey to a war-ravaged town called Kwangyingtang, where they managed to secure four pack mules and two litters of a type called "chair nests"—a rope basket attached to poles in which one rode suspended between two mules, protected from sun or rain by woven mats.

Leaving Kwangyingtang after being delayed by fighting and looting, Andrews, Collins, and their companions struggled along a road knee-deep in mud, passing wretched villages described by Andrews as "poverty-stricken and ravaged by cheap, readily available opium." Seven days later, they reached Sianfu—once the empire's capital during the Tang Dynasty (A.D. 618–907) and a place Andrews found "decidedly unfriendly." Here, they headed southward into the Tsinling Mountains, slowly ascending jagged, thickly forested cliffs and ridges. Andrews killed several prize examples of takin, but hampered by a sudden snowstorm, they abandoned the hunt. Andrews was overjoyed when two Chinese hunters he had left behind to wait out the storm arrived in Peking weeks later with three more magnificent specimens.

On the return trip, Andrews narrowly averted disaster when bandits invaded their camp. In exchange for safe passage, Andrews offered to treat the outlaws' wounds suffered in a scrimmage with soldiers of General Feng Yu-hsiang, the Christian warlord who was the *tuchun* or military governor of Shensi Province. Andrews' reliance on his medical skills, rudimentary as they were, no doubt saved his party from certain death at the hands of the roughly two hundred brigands ("Hard-looking fellows," Andrews wrote, ". . . and armed to the teeth") who straggled into the village of Lingtaimiao, where the hunters had occupied a deserted temple. After converting one room into a makeshift hospital, Andrews worked nonstop for hours on a stream of patients. "A chronicle of the wounds I dressed that day would not make pleasant reading. Suffice it to say that I learned how much of his anatomy a Chinese can lose and still carry on." When Andrews and Collins rode away on mules the following afternoon, they were escorted by five bandits, who saw them safely on their way to Sianfu. Grateful to have found the brigands such "decent fellows," Andrews later heard that when their food ran short at Lingtaimiao, they solved the problem by roasting villagers over hot coals.

Feng Yu-hsiang's behavior was hardly better than that of the brigands. A brutal tyrant despite his religious beliefs, he enjoyed a reputation for his disciplined troops and prowess on the battlefield, which he attributed to his conversion to Christianity while attending lectures at the YMCA in Peking. It was said that he went into battle inspired by biblical teachings, believing he was selected by God to cure China's ills. Feng required his troops to read the Bible every day. He opposed gambling, prostitution, alcoholic beverages, and the theater, denounced the wearing of silk as a sinful extravagance, and refused to ride in an automobile, preferring to travel about on a bicycle. Once he severely punished some of his officers for wearing boots he was sure no honest soldier could afford; and in his zeal to instill the virtues of economy, he instructed his men to conserve ammunition by executing prisoners with a single shot to the head rather than by firing squads.

At Sianfu, Feng summoned Andrews and Collins to berate them for "illegally" shooting takin, even though they had permits to hunt in Shensi from the minister of foreign affairs in Peking. Andrews loathed Feng on sight, and the interview with the *tuchun* quickly turned into a fiasco from which Andrews finally emerged victorious, refusing to be intimidated by Feng's outrageous threats. In a report of the incident, evidently prepared either for the United States legation in Peking or the War Department, Andrews described Feng as "abrupt, discourteous, and insulting," while Feng's opinion of Andrews as an "imperialist thug" was widely circulated throughout northern China.

In the spring of 1922, preparations for the Third Asiatic Expedition's journey into Mongolia rapidly intensified. Early in March, the geologist Charles P. Berkey, the photographer J. B. Shackelford, and the

chief of motor transport Bayard Colgate reached Peking, and the assistant geologist Frederick Morris came up from his teaching post at Pei
Yang University in Tientsin. Walter Granger had returned from his paleontological research in Szechwan, and by April 1, the entire staff was
assembled for the first time. Within a few days, the headquarters
seethed with activity. Tons of equipment were piled in the courtyards
and laboratory in various stages of packing. Each scientist and technician busied himself with checking every item needed to carry out his
specific duties. Before long, Colgate had transformed one of the courtyards into an open-air garage littered with tools, parts, and dismantled
Dodges and Fulton trucks. "Often the whirring of engines and clanking
of metal," wrote Andrews, "lasted well past midnight until Bayard was
satisfied that the motor cars were in perfect running condition."

While these activities were under way, Andrews' friend F. A. Larsen, acting as agent for the expedition, began purchasing the camels at
the equivalent of $40 each in U.S. currency. He also hired three Mon-
golian assistants and selected the camel drivers, whose leader, one of
Larsen's trusted employees, was a tough, resourceful nomad called
Merin, a man Andrews came to respect as "an incomparable master of
the desert." Everything to be transported by camel was packed into specially constructed wooden boxes. They were of two sizes, with sliding
tops, designed to accommodate supplies sent into the field each spring
and provide containers for specimens on the return trip. Every camel
carried four boxes—two large and two smaller ones—with a total
weight of about four hundred pounds; and because the two-humped
Bactrian camels native to eastern Asia shed their coats during the summer, the thick fur could be pulled off—amid pitiful whimpering and,
Andrews said, "large tears" streaming from their eyes—to provide packing material for fossils.

On March 21, the supply caravan, consisting of seventy-five camels
and six drivers led by Merin, left Kalgan and headed toward the Mongolian plateau. Andrews, who had come out from Peking to see them
off, instructed Merin to drop twelve cases of gasoline at a telegraph station called Iren Dabasu on the road to Urga, then proceed to the first
rendezvous with the automobiles carrying the scientists at Tuerin, a
lamasery located 150 miles from Urga. The fate of the expedition now
depended on Merin's ability to get the caravan to these resupply points,

even if it meant traveling at night to evade bandits. As Andrews watched the long line of camels depart for Mongolia's interior, its heavily armed drivers began chanting an eerie prayer to ward off evil spirits; and the lead camel, a huge, light-colored beast, sported a large American flag attached to one of its boxes.

With the caravan on its way five weeks ahead of them, the scientists rushed to complete their preparations. Everyone worked until late at night, except when attending the round of parties held in their honor. Two evenings before their departure, Albert B. Ruddock, the first secretary of the American legation, gave the explorers a gala farewell dinner. One of the guests, C. S. Liu, who was director of Chinese Railroads, became so intrigued with the expedition's objectives that he offered to ship the automobiles and equipment to Kalgan without charge and provide two private railway cars for the staff.

On April 17, the scientists, with their Chinese assistants, boarded the train and started for Kalgan amid a new flurry of rumors that Chang Tso-lin and Wu Pei-fu were about to fight it out for control of Peking. Using the Anderson, Meyer and Company's trading compound at Kalgan as a base of operations, equipment and personal baggage were loaded into the three Dodge touring cars and two Fulton trucks. To lighten the vehicles for the steep ascent to the Wanchuan Pass over a narrow, deeply rutted road, several hundred pounds of supplies, which failed to reach Peking from New York in time to be transported by the caravan, were piled into oxcarts and sent ahead to Miao Tan, about thirty miles beyond the Great Wall.

Andrews had arrived in Kalgan armed with "a formidable-looking document" from the Ministry of Foreign Affairs in Peking. It was supposed to permit the cars and equipment to leave Kalgan exempt from duty and customs inspection. But when Andrews presented it to Chang Tso-lin's soldiers, who were stationed along the road from Kalgan to the Wanchuan Pass, they laughed contemptuously. "This is from Peking," an insolent sixteen-year-old recruit snarled. "We don't recognize Peking." The soldiers demanded that another permit be obtained from the local military commander, something, Andrews protested, "that could have been resolved in ten minutes." It ended up requiring three days of telephone calls and meetings before the pass was issued.

At six o'clock on the morning of April 21, 1922, the five vehicles, "their engines roaring," Andrews remarked, "like the prehistoric monsters we had come to seek," drove out of Kalgan and climbed slowly westward toward the Great Wall. Joining them were two other cars, one driven by Andrews' wartime companion Charles L. Coltman, who was going to Urga on business. The other vehicle carried the wives of Walter Granger and J. B. Shackelford, who had come to China to see something of the country, together with Mrs. Davidson Black, whose husband, a professor of anatomy at Peking Union Medical College, was traveling with Andrews' party as far as Urga to conduct anthropological research.

When they reached the summit of Wanchuan Pass, the wives and their chauffeur turned back. Only Yvette continued on to Urga, intent on photographing a spectacular Lamaist festival—the Maidari—scheduled to take place on May 9. Other than this brief excursion by Yvette, no wives accompanied their husbands into the field by decree of the Museum's trustees. After considering the matter at a special meeting, the board had decided, as Osborn explained rather awkwardly in a letter to Andrews, that the presence of wives would cause "a division of responsibility and interest when every member should be entirely free to devote his thoughts as well as his time and energy to the expedition." As the project's leader, Andrews alone was given the choice of taking Yvette along for the entire season, but they opted against it: Yvette, after all, had the responsibility of a five-year-old son (who was being cared for by a nurse in Peking), and the dangers posed by the trip were too great.

Approaching the Great Wall, ascending through grass-covered hills and cultivated fields onto a plateau near the Wanchuan Pass, Andrews' euphoria intensified with each mile. Beyond lay Mongolia, with all of its mystery, challenges, and unanswered questions. "Wonderful panoramas were unfolded as we climbed higher," he wrote. "When we paused to cool the engines we looked back over a shadow-flecked badland basin, a chaos of ravines and gullies, to the purple mountains of the Shensi border. Above us loomed a rampart of basalt cliffs crowned with the Great Wall of China which stretched its serpentine length along the broken rim of the plateau. . . . The hills swept away in the far-flung, graceful lines of a vista so endless that we seemed to have reached the very summit of the earth. Never could there be a more satisfying entrance to a new country."

CHAPTER 10

Despite everyone's excitement at being under way, Andrews felt vaguely apprehensive as the cars drove toward the village of Miao Tan. Suddenly, the perils he had contemplated for so long were realities. Regardless of his careful planning, and his pronouncements that "adventures are a mark of incompetence," he knew perfectly well that a great deal now depended on luck.

At Miao Tan, the expedition paused long enough to load the extra equipment that arrived late from New York and was brought out from Kalgan by oxcarts. "When all these things were piled on the cars," Andrews said, "Colgate and I were horrified. There must have been at least two tons on each of the trucks, which were designed for only half that weight. There was no alternative, for the loads were made up largely of gasoline, photographic supplies, and automobile tires that could not be left behind."

Leaving Miao Tan at four o'clock, Andrews was anxious to reach the open grasslands to the northwest before nightfall, as the area through which the explorers were passing, a maze of hills and ravines, was a favorite haunt for bandits. Slowed by a drizzling rain, however, Andrews decided to camp at a Swedish mission at Hallong Osso, even though this meant driving after dark. But twenty-five miles short of their destination, the lead car, driven by Charles Coltman, suddenly sank up to its running boards in mud. One after another the rest of the cars slid off the rain-soaked road into a quagmire and had to be pushed out, although Coltman's car was so badly stuck it could only be extricated with a block and tackle. By the time the cars were safely on high ground, it was 1:30 in the morning and everyone was exhausted. Since proceeding to Hallong Osso was out of the question, the tents were hastily set up on a low hill. Because a suspicious band of heavily armed men, undoubtedly bandits, had ridden past earlier in the day, Granger, Shackelford, Morris, and Davidson Black were posted as lookouts.

At five o'clock on the following afternoon, the explorers camped in a beautiful grassy amphitheater enclosed by granite rocks, ate freshly killed antelope cooked over an *argul* (dried dung) fire, and cleaned mud from the previous night off equipment, clothing, and boots. En route to their next camping place at P'ang Kiang, a tiny outpost consisting of four earthen huts and a telegraph station, Berkey and Morris discovered eroded sedimentary strata that were ideal for fossil hunting, but Granger, who prospected the area all afternoon, failed to turn up anything of interest.

Twenty miles beyond P'ang Kiang, the travelers came to a Lamaist temple where they planned to refill their water bags at a nearby well. Always before on his trips to Urga, Andrews had looked forward to visiting this picturesque place, with its red- and gold-robed lamas who streamed across the plain on foot or horseback to greet him. But this time he found a horrifying sight. Recent fighting between Chinese and Russian troops had left the ancient temple deserted and partially wrecked by gunfire. As a ghastly reminder of the carnage that occurred at this once-serene spot, dozens of rotting soldiers' uniforms and lamas' robes were scattered on the plain and inside the empty buildings, some containing weathered human bones. Amid the deathly silence, the only signs of life were wild dogs lurking around the shattered temple, scavenging among the remains.

The grisly scene recalled all too vividly the tragic events that had engulfed Mongolia in the years from 1920 to 1922, the latest chapter in an historical legacy steeped in violence. For hundreds of years, barbaric tribes had waged constant warfare against each other for control of Mongolia's ill-defined territory. Not until the beginning of the thirteenth century were these disparate groups finally unified into the fabled empire ruled by Genghis Khan (1162–1227) and his descendants, which at its height stretched from China to eastern Europe and from India northward to Siberia and Russia. By the end of the seventeenth century, the Chinese had emerged as the dominant force in Mongolia; but with the fall of the Manchu Dynasty in 1911, the country's spiritual leader, the Living Buddha, or Hutukhtu, joined by hereditary nobles, issued a declaration of Outer Mongolia's independence. Only the area abutting China's western frontier, which came to be known as Inner Mongolia, remained permanently under Chinese control.

Yet China was deeply resentful over the weakening of its authority in Outer Mongolia. Ignoring the terms of a treaty signed at Kyakhta in 1915 by Russia, Mongolia, and China, which granted a measure of autonomy to Outer Mongolia, the Chinese established a sizable garrison at Urga in 1919 under the command of an unsavory tyrant named Hsu Shu-tung, who was better known as Little Hsu.

Inadvertently, China's illegal act opened the door for the appearance of Baron Roman Fyodorovich von Ungern-Sternberg, one of the most bizarre figures in Mongolian history. Known as the "Mad Baron," he was an enigmatic tyrant who might have stepped out of the pages of a Tolstoy novel. A tall, pale-skinned fanatic with red hair, withering blue eyes, a high-pitched voice, and a fearful saber scar across his forehead, he was ruthless, cruel, and ambitious. Descended from Baltic nobles who had entered the service of the czar, Ungern-Sternberg began his military career as a cadet in the Imperial Russian Navy before joining the Transbaikal Cossacks. Dismissed from the cavalry for incessant dueling and scandals involving officers' wives, he ended up in Mongolia as a dangerously paranoid alcoholic who survived by robbing caravans. Yet the country so captivated him that he converted to Buddhism, adopted a Mongolian name, and invested himself as the reincarnation of Genghis Khan.

Recalled to serve on the German front, Ungern-Sternberg fought with such reckless courage that he earned numerous medals and attained the rank of major-general at the age of thirty-three. In the aftermath of the Bolshevik Revolution, which he bitterly opposed, Ungern-Sternberg fled to Siberia in 1920. Within a few months, he had assembled a ragtag army of Cossacks, czarist White Guards, Japanese, Austro-Hungarian prisoners of war, Chinese, and Buriats—Mongols who lived in Siberia and held Russian citizenship. Obsessed by delusions of grandeur, he set out to conquer Mongolia, annihilate the Bolsheviks in Russia, and overrun Europe, thereby restoring the empire once created by his idol, Genghis Khan.

After two attempts were driven back by Little Hsu's forces, the baron, with an army of about six thousand troops, assaulted Urga on

February 1, 1921. Attacking from two directions, his forces broke through the Chinese defenses and began a frenzied slaughter of Little Hsu's troops, hundreds of which retreated into the desert toward China, vengefully killing every Mongol in their path. But rather than liberating the country from the Chinese, as many had expected, the baron, proclaiming himself military adviser to the Living Buddha and "God of War," unleashed a savage orgy of terror. According to a profile of Ungern-Sternberg in Fitzroy Maclean's book *To the Back of Beyond*, the horrors that engulfed Urga at the hands of the Mad Baron and his army defy description. Women were raped to death by gangs of soldiers; bakers were cooked alive in their ovens; elderly Tartars were made to dance for hours on rooftops in the bitter cold; Jews had water poured over their arms and legs until they were so solidly frozen they could be snapped off; Chinese prisoners were bayoneted in public or fed to the baron's pack of wolves; men, women, and children were shot, strangled, mutilated, hanged, burned alive, or lashed to death, while Ungern-Sternberg watched these atrocities with rapturous delight. Banks, shops, and private houses were looted: many were then set on fire, frequently with large numbers of people trapped inside.

Even as the baron reveled in victory, events outside Mongolia had already begun to seal his fate. On July 14, 1920, a young Mongol cavalry officer, Sukhe Bator, had ridden secretly from Urga to Irkutsk in Siberia and established contact with Lenin and other Communist officials. Originally, his mission was to enlist Russia's help in expelling the Chinese from his country. Since the baron had already driven out Little Hsu's oppressive regime, Sukhe Bator now asked the Red Army to intervene in eliminating the newest threat to Mongolia's sovereignty— Ungern-Sternberg and his rapacious troops. By then, Russia was openly desirous of extending Bolshevik control into Outer Mongolia, and the baron facilitated these ambitions by a fatal error in judgment: he decided to invade Russia, anticipating that huge numbers of peasants and soldiers, disenchanted with the Bolsheviks, would join his crusade. After several disastrous attempts to penetrate Russian territory, Ungern-Sternberg, his troops badly mauled and disillusioned, retreated into Mongolia, relentlessly pursued by the Red Army.

Once it became apparent that the situation was hopeless, his soldiers began deserting in droves, and soon thereafter the once-dreaded

Baron Ungern-Sternberg was seriously wounded by a group of his own officers in an abortive assassination attempt. Two days later, a Red Army patrol found him lying helplessly in the snow, nearly dead from loss of blood. He was taken prisoner, tried by a Russian military court, and unceremoniously executed by a firing squad.

Notwithstanding his ignominious demise, Ungern-Sternberg's ill-fated adventure was to have far-reaching consequences. Not only had the Chinese been decisively expelled from Outer Mongolia, but the country also was opened to powerful Soviet influences that resulted in the installation of a Communist government. Sukhe Bator returned to Urga in triumph, serving as minister of war and briefly as prime minister before his untimely death in 1923 at the age of thirty. Outer Mongolia officially changed its name to the Mongolian People's Republic in 1924, and thereafter the city of Urga was known as Ulan Bator (also spelled Ulaanbaatar or Ulaan Baatar), meaning "Red Hero," in honor of Sukhe Bator.

Against the background of these tumultuous events, Andrews proceeded farther into Mongolia, uncertain of the expedition's reception by the newly empowered government in Urga. But his apprehension was temporarily dispelled by an astonishing discovery. On April 24, four days after leaving Kalgan, Berkey, Morris, and Granger stopped to examine an area of sedimentary deposits. The rest of the party drove on to Iren Dabasu (which the Chinese called Erhlien), where Merin was supposed to have dropped off twelve cases of gasoline before continuing to the lamasery of Tuerin to await the cars. Upon reaching Iren Dabasu, whose only landmarks were a marshy lake and two mud houses, one of them a telegraph station, Andrews was relieved to find that the caravan had left the gasoline two weeks earlier. Shortly after six o'clock that evening, the tents were pitched on a ridge a half-mile to the west.

While Andrews and Yvette sat watching a fiery red-and-gold sunset, the cars carrying Granger and the geologists roared into camp. "We went out to meet them," Andrews recounted. "I knew something unusual had happened, for no one said a word. Granger's eyes were shining and he was puffing violently at a very odious pipe. . . . Silently he dug into his pocket and produced a handful of bone fragments; out of his shirt came a rhinoceros tooth, and the various folds of his upper garments yielded other fossils. Berkey and Morris were loaded in like manner. Granger held out his hand and said: 'Well, Roy, we've done it. The stuff is here. We picked up fifty pounds of bones in an hour.'"

Everyone began laughing, shouting, shaking hands, and pounding one another on the back. While dinner was being prepared, Granger wandered off to explore an outcropping and came back with half a dozen bits of fossils. Early the next morning, Berkey turned up two handfuls of bone, which Granger suspected were reptilian. A short time later, just as Andrews and Yvette were heading off on a shooting trip, Berkey summoned them to the top of a ridge, saying only that he, Granger, and Black had made an important discovery.

They found Granger on his knees, working gingerly with a camel's hair brush to expose the leg bone of what was unquestionably a *dinosaur*. It meant, Berkey said, that they were standing on the first Cretaceous stratum ever found in eastern Asia dating from the latter part of the Age of Reptiles (145 million to 65 million years ago); and the bone they were looking at was the first evidence of dinosaurs from this region. Moreover, the rhinoceros tooth and other fossils picked up earlier near Iren Dabasu came from the Cenozoic period, which began roughly 65 million years ago and had witnessed the emergence of mammals as the earth's dominant life form. Ecstatic over these revelations, Andrews boldly declared, "[We] had opened up a paleontological vista dazzling in its brilliance. With the [mammalian fossils] that had been found the day before, the dinosaur bone was the first indication that the theory upon which we had organized the expedition might be true: that Asia is the mother of the life of Europe and America.

". . . All this in a region we had been told would yield us nothing! Can one be surprised that we were jubilant?" So jubilant, in fact, that even a sudden dust storm failed to diminish the explorers' excitement— a wind that struck with such force it knocked down and ripped tents, scattered equipment and clothing over the plain, and sent a Standard Oil tin in which the Chinese cook, Liu, was baking a goose hurtling into the sand.

Leaving Granger, Berkey, and Morris to continue searching for more fossils, Andrews and the rest of the party departed on the 250-mile trip to the Tuerin Lamasery, where they planned to meet Merin and the supply caravan. With everyone crowded into three vehicles (two were left with Granger and the geologists), Andrews started on April 25, crossing the border between Inner and Outer Mongolia at a remote telegraph station called Ude. Not far into the journey, Andrews' party came upon more chilling evidence of the violence that had swept over these plains during the previous winter. Along the roadside were the remains of perhaps a hundred Chinese soldiers, still swathed in sheepskin clothing, who had apparently been caught in a blizzard and frozen to death during their frantic attempt to escape Ungern-Sternberg's Cossacks. Many of the corpses had been picked clean by wolves or vultures, but despite the gruesomeness of the scene, Davidson Black rummaged through the bones collecting skeletons for his anatomical laboratory at the Peking Union Medical College.

Just before reaching Tuerin, Andrews sighted a caravan camped beside a pond. He instantly recognized his boxes and saw an American flag fluttering above one of the tents! Merin, grinning broadly and waving, rode out on a camel to meet the cars, announcing proudly that he and his drivers had arrived at Tuerin *one hour* ahead of the automobiles, with all seventy-five camels and their loads intact. "I had such a feeling of satisfaction," Andrews confided in his journal that night, "as seldom comes to a man."

Moving the cars to a telegraph station near the Tuerin Lamasery, Andrews directed that a camp be set up among some sheltering rocks not far from a deep well. He then sent Colgate to guide Merin and the caravan to the camp, while Shackelford positioned his Akeley camera to film their arrival. "About half past three," Andrews recorded in his journal, "Shackelford yelled, 'the camels are coming' and we dashed out of our tents." In the distance, the lead camel, its flag whipped by a gentle breeze, plodded toward the camp, followed by the others in single file. "My blood thrilled at the sight," wrote Andrews, ". . . it impressed upon me, as nothing else had, that the expedition was a reality. The camels swung past the tents, broke into three lines like files of soldiers and knelt to have their loads removed; then, with the usual screams of protest, they scrambled to their feet and wandered down the hill-tops to the plain, nibbling at the vegetation as they went."

So far, the expedition had justified Andrews' most optimistic expectations. His plan to enable the scientists to function as independent units seemed to be working; the caravan had kept its rendezvous on schedule; and the fossils discovered near Iren Dabasu promised greater scientific rewards to come. Ahead of him, however, were the political uncertainties he faced in Urga, the complexities of which Andrews had no way of gauging. He had expected to telegraph F. A. Larsen from Tuerin regarding his upcoming negotiations with Mongolian authorities, but the line had been cut beyond this point, apparently by Chinese soldiers fleeing Ungern-Sternberg's forces. During the blood-drenched winter of 1921, a massacre had occurred near the telegraph station and in the rocky escarpments around the Tuerin Lamasery when four thousand luckless Chinese were virtually annihilated by three hundred Mongols led by a legendary Buriat general, whom Andrews later met in Urga. Even now, empty rifle shells, cartridge clips, and pieces of uniforms littering the ground bore grim testimony to the terrible slaughter. Andrews heard eyewitness accounts from the lamas, who had watched in horror from behind the walls of their lamasery as the demoralized Chinese troops were taken by surprise in a predawn attack and shot at point-blank range, cut down with sabers, or clubbed to death with rifle butts almost to the last man.

Andrews' disappointment at finding no telegraph service between Tuerin and Urga was quickly dispelled by a letter waiting at the station addressed to "Roy Chapman Andrews, Esq., Anywhere in Mongolia." Aware that the expedition was traveling along the Kalgan-Urga road, the letter had been carried from Urga by a telegraph company agent sent to repair the line. It was from Larsen, who reported that conditions appeared favorable for working out an agreement with Mongolian officials and urged him to hasten to Urga. Instructing Merin to await the arrival of Berkey, Granger, and Morris from Iren Dabasu, Andrews departed on May 2 with Yvette, Shackelford, Colgate, and Davidson Black.

Driving into Urga that afternoon, they found a city in violent transition. No longer the free-spirited place last seen by Andrews in 1919, it had become the strictly regimented capital of a Communist state. Outwardly one still saw vestiges of Urga's mesmerizing blend of color, sounds, and smells. But this first impression was deceptive, and Andrews was shocked by the changes that had beset Urga. "Now there were not so many dashing horsemen on the street," he observed, "not so many strange costumes and half-dazed nomads from the steppes of Tibet or the deserts of Turkestan. Their places have been taken by Russians and swaggering Buriats, and the great open square is filled every day with squads of awkward Mongols being drilled as soldiers. Machine guns stand where lines of camels knelt before, and the prison is filled to overflowing."

Gone, too, was the ease with which foreigners once visited Urga. Anyone entering or leaving was automatically questioned, their papers examined, and all luggage searched. "In short," Andrews complained, "one was treated as a spy and a generally undesirable character. With reluctance the authorities might admit at first sight there was no reason to detain the traveler, but from the moment he arrived in Urga until he departed, spies were detailed to watch and report his every movement. Woe to him who did not destroy every scrap of written or printed matter before he fell into the clutches of the Secret Police; even carrying the best American magazines could land him in jail for a night until the authorities had perused them in the hope of finding 'seditious literature.'"

Immediately upon reaching Urga's outskirts, Andrews' cars were stopped by Mongolian soldiers—members of Sukhe Bator's new national army—who inspected the temporary entry permits issued at Ude, the border crossing from Inner Mongolia. Three soldiers then jumped on the running boards and escorted the travelers to a customs office, where their permits were examined again and stamped. At yet another customs station closer to the main square their baggage, sleeping rolls, and equipment were unpacked for inspection before everyone climbed back into the cars and proceeded to the military commandant's headquarters. As he had gone home for the night, the usual interrogation to which foreigners were subjected was postponed until the next day. Retreating to Larsen's house for a hot bath and dinner, Andrews spent a restless night contemplating the nerve-racking ordeals he was certain awaited him.

Early the following morning, his worst fears began to unfold. "While at breakfast," he recalled, "a soldier appeared and asked us to come to the military [headquarters]. There they told us they would send us our answer in the p.m.—Answer to what???" And so it went through one episode after another over the next two weeks, with gov-

ernment officials indulging in a Kafkaesque scenario of irrational, arbitrary, and sometimes comical posturing and red tape. Andrews soon realized that his expedition was deliberately being used as a pawn in a
power struggle between Mongolian authorities, not yet fully indoctri-
nated into Bolshevik ideology, and Russian and Buriat officials bent on
dismantling the old regime as rapidly as possible. Because the respected
and well-connected Larsen—aided by his close friend T. Badmajapoff,
an adviser to the minister of justice and, ironically, a Buriat—had been
diligently working behind the scenes on Andrews' behalf, the Mongolians were inclined to approve the expedition's plans. Infuriated by this
prospect, the Russians and Buriats, who were suspicious of Andrews'
objectives, sought to convince the Mongols that his expedition was actually a capitalist ploy. With fervent anti-Communists and "robber
barons" like Morgan, Dodge, Rockefeller, Baker, Harriman, and James
among the project's financial backers, the Soviets argued that the explorers had come to Mongolia under the guise of science to spy and
prospect for oil and minerals, which American and European interests
would ruthlessly attempt to exploit. Nor did it help Andrews' position
that the Standard Oil Company had donated three thousand gallons of
gasoline and fifty gallons of oil to the expedition.

On the morning of May 3, Andrews was told by the minister of foreign affairs that the Russians had ordered the supply caravan brought to
Urga; it was to be inspected by customs officers and duty levied on the
camels and their loads. Later the same day, after Badmajapoff inter-
ceded, the minister decreed that the caravan could remain at Tuerin.
Next came word that the Russians were insisting on sending two representatives to Tuerin to examine the caravan. They also planned to interview and photograph the other members of the expedition's staff,
including all Chinese technicians and servants, as well as the Mongol
camel drivers. "Utter rot," Andrews responded angrily, ". . . just one
more thing the Reds are thinking up to make trouble." Nevertheless,
Bayard Colgate and Badmajapoff dutifully drove the two officials the
150 miles to Tuerin, accompanied at the last minute by "a little Russian
Secret Service man," wrote Andrews, "because [the Russian consul]
feared the presence of spies among our staff."

Five days later, Badmajapoff informed Andrews that the Soviets
had compiled a printed form containing twenty questions involving the

name, age, nationality, and other personal information pertaining to every member of the expedition which was to be submitted, along with two photographs of each man, to the passport office in Urga. Andrews promptly dispatched a car driven by Charles Coltman's chauffeur to deliver the questionnaires to Colgate at Tuerin. He secretly included a note telling Colgate that as soon as Granger and the geologists returned from Iren Dabasu, he should prepare to move the expedition to a place eighteen miles west of Urga called Bolkuk Gol.

Andrews then learned that a contract was being drawn up for his signature that set forth stringent rules governing the expedition's activities. Even worse, a Russian Secret Service agent had been assigned to accompany the explorers to ensure that they abided by the terms of the contract. Andrews exploded! "If they insist on sending him, as they undoubtedly will, he is going to have a beautiful little trip on a camel for I won't have the wretch in my cars. . . . Not only are [the Reds] going to force me to take the man, but also to furnish food, transportation, & living quarters—next the devils will insist that I pay him a salary!" And that was not the worst of it. Having a Secret Service agent underfoot meant the complete disruption of Andrews' plans to limit the first season's activity to a reconnaissance aimed at locating scientifically promising areas for future investigation. "They could not have more successfully thrown a wrench into my work," Andrews fumed. "It will force me to change all my plans for the summer because if we find fossil fields we will have to stop and [excavate] them . . . or else run the certain risk of having the Russians send an expedition & steal all the results of our work." Andrews implored Badmajapoff to intervene with the Russian consul, a scoundrel named Ochten, who Andrews suspected was behind the idea of sending along the agent. But Badmajapoff felt there was no way out. If Andrews refused to take the man, he believed, Ochten would surely say to the Mongols, "You see, the Americans have something to hide after all."

While waiting for a resolution to these bizarre complications, Andrews, Yvette, and Shackelford passed the time by photographing Urga, visiting temples, perusing marketplaces and shops, and socializing with the dwindling number of foreigners who still lived in the city. On May 9, Urga's population turned out en masse for the Festival of the Maidari, or "Coming Buddha," one of Mongolia's most important religious events, which Yvette had expressly come to Urga to photograph. Early in the morning a gilded image of the Maidari—a highly venerated bodhisattva*—was taken from a temple in which it reposed, placed on a huge throne festooned with prayer flags and streamers, and drawn through the streets by an elaborate procession of worshipers, carrying, umbrellas to shield them from the blazing sunlight.

"At ten o'clock, when we reached the main square," Andrews wrote, the procession had not yet appeared, but the air was throbbing to the boom of drums and the deep notes of conch shells. Suddenly, a great mass of color could be seen advancing from the east. Every shade of the spectrum was repeated a hundred times in the gorgeous pageant of lamas, Andrews observed. As the procession moved closer, he recognized the premier wearing a robe spun of gold with a sable hat on his head. Beside him were the country's four reigning khans of Mongolia (hereditary nobles who claimed direct descent from Genghis Khan), and behind them marched a double row of princes, dukes, and lesser nobles dressed in dark blue gowns with brilliant cuffs and embellished with peacock plumes.

The lavish throne bearing the Maidari was shaded by a silk canopy of rainbow colors and surrounded by the highest-ranking lamas, resplendent in gold cloth. From the throne, silk ropes were held by lamas who flanked the sacred image on all sides and carried huge umbrellas of red and yellow silk. Behind the Maidari were throngs of other lamas—almost ten thousand of them—and two or three thousand men, women, and children followed in a chaotic mass. Most of the women were of noble birth; they wore rich gowns with ropes of pearls about their necks and hair ornaments of gold studded with precious stones.

Reaching the Gandan Lamasery, Mongolia's holiest temple, the crowd halted. High lamas seated themselves on colorful prayer mats, with the premier in the center and a group of gorgeously bedecked lamas surrounding the Maidari. Andrews noticed that the seated priests

---

*A bodhisattva is a being who compassionately refrains from entering nirvana in order to save others. Bodhisattvas are worshiped as deities in Mahayana Buddhism.

were served tea and food while a red-robed lama standing on the Maidari's throne energetically thumped the heads of the populace with a long stick padded at the end. "There could not be the slightest doubt in the mind of the suppliant that he had been blessed after the ball [on the] stick landed on his head, for at times the officiating lama took huge delight in bringing it down with force enough to rock his victim."

By mounting his motion picture camera on a watchtower above the square, Shackelford was able to film the ceremony from an ideal vantage point. Meanwhile, Yvette, moving through the crowd, took dozens of color photographs. As it turned out, the pictorial record of the Maidari captured by Yvette and Shackelford assumed an unanticipated historical importance. No one realized it at the time, but this was to be the last Maidari Festival ever enacted. With the death of the Hutukhtu the following year, and a concerted effort by the Communists to rid Mongolia of Lamaist Buddhism, the Maidari was permanently banned, along with all other public religious ceremonies.

Early in the afternoon on May 10, Yvette began her return trip to Peking. Traveling with two cars, she was accompanied by Davidson Black, a Swedish diplomat identified only as Dr. Essen, and an Anderson, Meyer and Company employee and his wife, the Hansens. Also included among the group were four Chinese passengers and one of Larsen's Mongol assistants, whom Andrews was sending to Tuerin to lead the expedition to the new rendezvous point at Bolkuk Gol, west of Urga.

Andrews' parting with Yvette was agonizing. War clouds looming over China and the uncertainties of the expedition enveloped them in a disquieting sense of apprehension. Everything about their future now depended upon factors beyond their control. Yvette's anxiety was intensified by a growing fear that the expedition was threatening her marriage. Obsessed with realizing his dream, Andrews had succumbed completely to its relentless demands on his time and energy. Increasingly, it seemed, there was less room for anything else in his life, and his divided loyalties were poignantly illustrated by a journal entry written the day Yvette left Urga: "The last I saw of Yvette was her little white handkerchief waving me goodbye. Five months is a long time and much may happen to us both. There is a war in China but I think she will be safe. . . . No telling what will happen to me—if anything does happen—if any of the expedition doesn't come back—I hope it will be me. The confidence the men place in my judgement and my ability to bring them back safely, makes me realize more and more what a responsibility I have."

The next day, a traveler arrived from China with word that the long-expected clash between Wu Pei-fu and Chang Tso-lin had finally erupted. It ended in a stunning defeat for Chang's troops, and the victorious Wu Pei-fu was preparing to march on Peking at any moment. More bad news came on May 12, when Larsen delivered a letter sent to Andrews by Yvette from Tuerin informing him that the automobile carrying the Chinese had broken a wheel and turned over twice. All the passengers had suffered serious injuries, and the Mongol assigned by Andrews to guide the expedition to Bolkuk Gol had received a fractured skull and a broken collarbone. Although Yvette's car was not involved, the accident left Andrews so shaken he was unable to sleep for several nights. In his journal, he admitted feeling "half sick . . . thinking of 'what might have been'—I can't get it out of my mind. What a relief it will be to know that Yvette is safely in Peking."

As far as obtaining permits for the expedition, there was no choice but to wait. "All foreigners in Urga are simply waiting," Andrews commented with resignation. "They go about from one house to another & talk to pass away the time. Everyone is waiting for official permission to do something or other. It would be comic if it were not so annoying." Outsiders invariably found themselves inundated by regulations. Automobiles were not allowed in the streets after dark without a permit. Using a camera required special permission. Visitors wishing to go sight-seeing usually were accompanied by police. Foreign-owned shops and businesses were disappearing rapidly because of bureaucratic restrictions and taxes, and exorbitant duty was charged on all goods entering or leaving the city—except for items traded with Russia. "If this is a sample of Bolshevik rule," Andrews grumbled, "deliver me from any further contact with it."

Andrews conceded, however, that one badly needed reform had occurred in Urga since his last visit in 1919. Its unspeakably cruel prison had been replaced with a vastly improved facility. Gone were the stacks of cramped wooden boxes in which inmates, unable to lie down or sit upright, served out their sentences in unimaginable agony, given food through a small hole and forced to endure their own excrement. Instead, the new prison consisted of four large buildings laid out around an inner courtyard. Inmates lived in twelve-by-eight-foot cells, and all of them, Andrews reported, seemed well fed and clothed, with access to the courtyard for exercise. But Andrews refused to admit that the Communists might have been responsible for getting rid of the dreaded "torture boxes." "I should hesitate," he declared, "to credit *anything* of progress to them from what I have seen."

Nevertheless, he unabashedly approved of the Communists' determination to purge Mongolia of Lamaist Buddhism. Introduced from Tibet during the reign of Kublai Khan in the thirteenth century, Lamaism was a mixture of Mahayana Buddhism infused with shamanism. It rapidly became the country's official religion, and its spiritual leader—known variously as the Hutukhtu and Bogdo Gegen or Living Buddha—ruled the country from his seat of power in Urga. His control over religious and secular affairs was absolute until the winds of revolution began to sweep Mongolia early in the twentieth century.

Andrews loathed the influence of Lamaism, believing that it led to what he termed "the decadence of the Mongolian race." Because its teachings required that one son born to every couple must become a lama—and it was not uncommon for all boys in a family to join the sect—at least one-third of Mongolia's male population donned the red and gold robes worn by the priesthood.

Sequestered in dozens of lamaseries, ridden with syphilis, and supported entirely by tribute, thousands of supposedly celibate lamas "who were little more than unproductive beggars," Andrews wrote, "spent their time chanting Tibetan prayers they did not understand and living as dissolute human parasites whose beliefs fostered superstition and discouraged learning, enterprise, and ambition." Nor was Andrews alone in this evaluation of lamas: many other travelers also shared his scathing indictment of Lamaism, vigorously denouncing it as hopelessly mired in ignorance and arcane ritual.

In their zeal to eradicate Lamaism, the Communists resorted to drastic measures. Most of the older lamas, together with any who resisted the government's policies, were executed, and the younger, ablebodied men were forced into vocational training programs. With a few exceptions, the lamaseries—over seven hundred in all, some of which were architectural landmarks—were either blown up or abandoned to ruin. Officially, the constitution of the Mongolian People's Republic guaranteed freedom of worship, but a law was passed decreeing that no male could become a lama without first completing eight years of compulsory education, during which students were made to study the evils of Lamaism. As a result, few wished to enter the priesthood, and not until the Russians withdrew from Outer Mongolia in 1990 did Lamaism undergo something of a revival.

Exactly two weeks after Andrews had arrived in Urga, a break in his rancorous diplomatic negotiations finally came. At nine o'clock on the morning of May 16, Badmajapoff delivered a copy of the government's contract defining the conditions under which the expedition could operate. After reading it, startled to find most of its provisions quite lenient, Andrews amended several unacceptable points. Early that afternoon, he and Larsen appeared at the Foreign Office for a meeting with the premier, the minister of foreign affairs, and a group of advisers who had "gathered in a solemn conclave." Sensing at once that the Mongolian officials, heavily swayed by Larsen and Badmajapoff, had outmaneuvered the Russians in their opposition to the expedition, Andrews seized the initiative. Speaking in Chinese and directing his remarks almost exclusively to the minister of foreign affairs, he boldly demanded that the contract's objectionable points be deleted, which the minister agreed to do without protest.

In the end, Andrews came away from the meeting with everything he wanted. The contract prohibited the explorers from making "topographical sketches" (a vague restriction subject to broad interpretation) or prospecting for minerals "by way of boring or deep digging." Otherwise, they were free to travel wherever they pleased, to collect fossils, zoological and botanical specimens, and rock samples, to take photographs and motion pictures, and compile maps. In return, Andrews promised to supply the Mongolian government with duplicates of everything they collected, and to furnish copies of all maps, photographs, moving pictures, and publications relating to the expedition's discoveries.

Inexplicably, when the touchy subject of the Russian Secret Service agent who was supposed to accompany the expedition was raised, it was agreed, or so Andrews wrote in his journal, that except for flour and fresh game, he would have to provide his own food and live with the Mongol camel drivers. In his summary opus on the expeditions, *The New Conquest of Central Asia*, Andrews relates that the contract obligated him to "take a government official with us to see that we carried out our arrangement." In reality, however, the idea, purportedly concocted by the nefarious Russian consul, Ochten, to send along one of his handpicked spies, evaporated. "The official designated to go was our friend Badmajapoff," Andrews stated, "with whom it had all been arranged previously. Badmajapoff is a Buriat who has had considerable experience with the great explorer P. K. Kozloff. His presence on the expedition was most agreeable to us all and made for us a firm friend, who in later years did much to help me steer a safe path among . . . the political rocks which barred the way to the great open spaces of the Gobi Desert." Andrews had also persuaded the indefatigable Larsen to join the expedition as an interpreter, aware that his presence would be equally valuable in reassuring the government of the expedition's integrity.

Andrews described the meeting at which the contract was ironed out as "too comical for words. It was as solemn as tho' we were deciding the fate of nations." When the formalities had ended, the grave-faced premier asked Andrews to grant the government a special favor: would he attempt to collect a specimen of the *allergorhai-horhai*, a legendary creature reportedly encountered by countless nomads in the desert? Shaped like a sausage, about two feet long, it had no head or legs, and was supposedly so poisonous that merely touching it meant instant death. Andrews assured the premier and other ministers, some of whom had friends or relatives who claimed to have seen the beast, that he would make every effort to bring back a live example. With thinly disguised humor, totally lost on his deadly serious audience, he suggested that wearing dark glasses might minimize the disastrous effects said to result from looking directly at the monster, and that perhaps it could be safely caught using long steel forceps.

Twenty-four hours earlier, the explorers' passports had been issued, along with permits to carry hunting rifles, shotguns, and pistols. A waiver was also granted exempting their automobiles, camels, equipment, and personal belongings from duty. Word had been received from Colgate that Granger, Berkey, and Morris, after finding rich fossil deposits at Iren Dabasu, had arrived at Tuerin and rejoined the expedition, which was making its way to Bolkuk Gol. Elated by this news and the lifting of political obstacles in Urga, Andrews, Shackelford, Larsen, and Badmajapoff departed the city on May 18, agreeing regretfully that Urga was "an exceedingly good place to leave."

Driving over a bridge across the Tola River, the cars headed westward to join the rest of the expedition. "Nothing ever looked so good to me," Andrews recounted, "than the blue tents pitched on the side of a gentle slope with a great snow-bank glistening in the distance. Beyond was a small temple surrounded by half a dozen Mongol yurts; the place rejoiced in the name of Bolkuk Gol.

"Not more than two hours after we reached camp, Merin came galloping in on his great white camel. He reported that the caravan was only half a mile away and that all the animals were in good condition. Soon we saw the long line of camels silhouetted on the summit of a hill with the American flag streaming above the leader. . . .

"We celebrated our reunion with a huge dinner, and I went to sleep with peace and thanksgiving in my heart. The last barrier had been passed and before us lay an open trail."

PART THREE
Into the "Great Unknown"

Somewhere in the depths of that vast, silent desert lay those records of the past that I had come to seek.

—Roy Chapman Andrews

CHAPTER 11

The Mongolian People's Republic—still known as Outer Mongolia in 1922—encompasses an area of approximately 604,800 square miles. Shaped rather like an elongated oval, it extends for about 782 miles from north to south and 1,486 miles from east to west. Buried deep within the heart of east-central Asia, totally encircled by Russia and China, Mongolia comprises a basinlike plateau, nearly all of which lies at an elevation of three to five thousand feet above sea level. Two major mountain ranges punctuate its landscape: the Khangai in north-central Mongolia, whose peaks vary from four to over twelve thousand feet in height, and the rugged Altai Mountains, which sweep down from the country's northwestern border and thrust southeastward for a thousand miles, reaching a maximum elevation of 14,350 feet at Nayramadlïn (or Hüyten) Peak at the far-western edge of the Altai range.

Immense sections of northern Mongolia are covered by forested steppe and lush grasslands drained by swiftly flowing rivers that empty into hundreds of lakes. The region's largest river, the Selenge (or Selenga) and its principal tributary, the Orhon, flow northward across the Russian border and drain into Lake Baikal. Farther south and west the terrain becomes increasingly arid as it merges with the Gobi. Second in size only to the Sahara among the world's deserts, the Gobi blankets southern Mongolia and stretches into the Chinese provinces of Sinkiang, Kansu, and Inner Mongolia. Entering the Gobi's overpowering vastness, the traveler encounters an awesome landscape: eroded badlands, salt flats and marshes, rocky outcroppings, brackish lakes (which often disappear in dry years), sand dunes, escarpments, the barren peaks of the Altai Mountains, and seemingly illimitable expanses of gravel-covered plains known as *gobi*, a Mongol word from which the desert acquired its name.

The country's climate is unpredictable and severe. Annual rainfall in the northern highlands measures around fourteen inches, and the Gobi generally receives from three to eight inches. Winds continually blow across the Mongolian plateau, and raging dust storms plague the desert, whipping sand and gravel through the air with stinging velocity. Winters usually last from October into May, bringing heavy snowfall and temperatures of minus 50 degrees or below (Fahrenheit) in the northern districts. Even the Gobi is subject to blasts of Arctic cold and deep snow, a fact that caused Andrews to withdraw his expeditions no later than mid-October every year. Nor are summers in the desert immune to climatic extremes: it is commonplace for sweltering days, often climbing to 120 degrees or more, to alternate with bitterly cold nights or occasional snowfall.

Unlike in most deserts, wells in the Gobi were seldom difficult to locate. Andrews noted that the water table was rarely more than twelve to fifteen feet below the surface, especially in dry streambeds or where inclining geologic strata had channeled drainage to low points. Adequate wells, some quite old and lined with stones or sod, occurred every ten to sixty miles along regularly traveled caravan routes. "Mercifully," Andrews remarked, "the best of these provided sufficient water not only for drinking, cooking, and the automobiles' radiators, but for occasional much needed baths."

Although Andrews' frequent references to the Gobi as the "Great Unknown" was a provocatively romantic metaphor, it was somewhat misleading. For centuries, a handful of travelers had ventured into Mongolia's uninviting wastes. As early as 1224, the famous Venetian wayfarer, Marco Polo, had traversed the southern Gobi on his way through China. Not long afterward, two Franciscans, John de Plano Carpini and William of Rubruck, visited Genghis Khan's legendary capital of Karakorum in north-central Mongolia. Friar William, who reached the city in 1253 as an emissary of King Louis IX of France, recorded vivid accounts of Mongolian society under the reign of Möngke, one of Genghis Khan's grandsons.

Beginning in the seventeenth century, a succession of travelers and Jesuit missionaries, most of them bound for China, passed through

Mongolia. But it was not until the nineteenth century that the country attracted serious scholarly interest, awakened mainly by fabulous archaeological wonders unearthed in the sand-covered cities that once flourished along the Silk Road, which served for centuries as the main artery for richly laden caravans traveling back and forth between the Mediterranean and China. By the 1870s, a series of expeditions—treasure hunts, in reality—was undertaken by Russian and European explorers, most notably Nikolai Prejevalsky, P. K. Kozloff, Sir Aurel Stein, Raphael Pumpelly, Ferdinand von Richthofen, V. A. Obrechev, Francis Younghusband, and Sven Hedin. As these explorers were primarily archaeologists, their observations of botany, geology, and zoology were haphazard at best. No attempt had ever been made to apply the kind of multidisciplinary approach conceived by Andrews. And, to be sure, there were still enormous areas of the Gobi in which no explorer had ever set foot.

On May 19, Andrews decided to proceed to a lamasery called Sain Noin Khan, about three hundred miles to the southwest. "It was a leap in the dark," he admitted, "but if the country did not give us results we could always turn south into the Gobi." After five days, the explorers reached another lamasery, Tse Tsen Wang, named for a prince who ruled over the province in which the settlement was situated. Andrews described it as a "miniature city of tiny houses, temples and pinnacled shrines, surrounded by enormous piles of argul." Tse Tsen Wang, one of Mongolia's richest and most powerful princes, was an old friend of Larsen and Badmajapoff; Andrews had first met him in Urga in 1919 and purchased several horses from him. About two miles beyond the lamasery, the expedition pitched its tents in a sheltered canyon near an ice-cold spring. The party remained here for a week awaiting the caravan, which they learned from a traveling lama had been delayed by the weakened condition of the camels due to unexpectedly poor grazing along the way. Adding to the men's apprehension, Andrews was now forced to spend long periods confined to his tent, suffering from an acute eye infection, almost surely ophthalmia caused by exposure to the desert's intense sunlight. Resting his eyes in the shade and applying wet compresses temporarily eased the discomfort, but the condition gradually worsened, resulting in permanent damage to his vision.

The caravan finally appeared on May 28, only slightly the worse for the camels' lack of food. Two days later, the expedition departed for Sain Noin Khan, another 150 miles farther south. Each sun-parched mile carried the men deeper into Mongolia's enveloping void. As the country became more desolate, the scientists began encountering nomads who had never seen white men or automobiles and who fled in terror at their approach. It sometimes took hours to convince the natives that the mysterious *chi chur*, or "wind carts," as the Chinese called the vehicles, were harmless. Often the sight of the cars caused the men to panic, leap on their horses, and gallop away at breakneck speed, without regard for their families. On one occasion, three elderly women and a beautiful girl of eighteen or twenty were instantly abandoned by the men at the approach of Andrews' car. Trembling with fright and certain of being killed, the women spread a piece of white felt in front of their yurt and lined up around it, awaiting their fate. Even after the Mongol interpreter assured them the visitors meant no harm, the young girl brought out cheese and tea in an attempt to placate the strangers. Once, when Granger and Colgate stopped at a yurt, a terrified woman—the men having dashed away on horseback— became so nauseated from fright that she vomited uncontrollably, prompting Shackelford to comment that the Mongols' creed might well have been, "Save the men, to hell with the women and children."

Adding to the explorers' mounting sense of isolation, their wireless receiver failed to operate. Before leaving Peking, Andrews had arranged with the American legation to beam news every evening at seven o'clock, along with time signals for use by the geologists in checking their chronometers to determine latitudes and longitudes. "We had purchased the receiving set in Peking," Andrews said, "[and] it looked very business-like when the aerial was erected on tent poles . . . but Shackelford and Colgate could not get a sound over the wire. . . . After we left Urga, we were completely without news, although the legation sent out messages every night during the five months we were away."

More disturbing were persistent reports of bandits in the region the expedition was approaching, especially one group of over a thousand brigands operating in the southwestern Gobi. "They were under the command of a well-known chief," Andrews learned, "who had declared war upon all Russians and adherents of the Soviet-controlled Mongol government. Any captured Russians—and that phrase really meant white men, since to the Mongols all white men were Russians—were tortured in the most inhumane way. One man was skinned alive."

By now, too, Andrews was beginning to despair regarding Yvette's safety and the political situation in Peking. He had heard nothing since her letter of May 12 in which she reported the automobile accident that delayed her party at Tuerin. In one of the rare introspective passages in his journals, Andrews poured out his concern:

> I have been very depressed the last two days for I want to hear whether Yvette is safe in Peking or not. . . . Is it all worthwhile? I wonder! Certainly an explorer should never marry—a wife and children do not go well with the life he must lead. Yet other men have done it—Peary, Scott, Shackleton all had wives and children. Perhaps I am not as strong as they are—if only I knew that Yvette was safe & well I think I could rise above the loneliness but it is a hard load to carry when I know that I must go for months yet with the terrible uncertainty. . . . If only our wireless worked so that we could get news.

As the procession of vehicles made its way southward, the country changed from gently rolling hills into what Frederick Morris called "a geological nightmare," a ragged landscape of boulders, granite outcroppings, and stony dikes thrusting upward from the surface. Even the few flat spaces were dotted with obstructions known as "playa tuffs," formed by sand caught in the roots and lower branches of desert vegetation, creating oddly shaped mounds from one to eight feet high. Just beyond these badlands lay another stretch of verdant, undulating plains through which a shallow river, the Ongin Gol, meandered in a winding course. Despite an abundance of water and grass, ideal for grazing, there were no yurts or animals anywhere in sight, and an unearthly silence was broken only occasionally by the wind. When Andrews sent one of his Mongol assistants, Tserin, to a nearby ridge to shoot sand grouse for the evening's meal, the reason for the absence of nomads became evident. Tserin returned with two terrified natives who told Andrews of a gruesome incident:

Only a few months earlier there had been an encampment of seventeen yurts almost where our tents were pitched. It was a peaceful Mongol village numbering about fifty individuals. One day a Chinese caravan from Uliassutai, on the way to Kweih-wating, laden with sable skins and other valuable furs, had camped beside them. During the night a party of Russians arrived and slaughtered every living soul, men, women, and children. . . . So far as they knew, not a human being was left alive to tell the tale. Then the Russians drove off the camels with their load of furs, and all the Mongols' sheep and goats. Even the dogs were shot before they left.

By June 3, the expedition came within view of Sain Noin Khan, which sprawled in a broad green valley, "an indescribably beautiful sight," Andrews remembered, "for its temple roofs, golden spires & upturned gables glistened in every color of the rainbow." Flanking the compound on two sides were wooden dwellings that housed over a thousand lamas. The entire city was encircled by a palisade made of unpeeled logs, and scattered among the surrounding hills were religious shrines known as *obos*, piles of stones erected by nomads or pilgrims, containing prayer flags, offerings of money, and religious talismans placed among the rocks.

Although the prince in whose ancestral domain Sain Noin Khan was located was only ten years old and residing elsewhere at the time, the young noble's regent, a high lama, was temporarily ensconced at a sacred mountain known as Arishan, a few miles from the lamasery. Near its base was a shrine venerated for a hot spring that issued from under a ledge, pouring forth water that was believed to possess healing properties and drawing crowds of pilgrims. In the hills above Sain Noin Khan, Andrews established an idyllic camp amid a thick larch forest. For two weeks, a constant stream of nomads passed through the valley below on their annual migration from winter camps in the south to the northern grasslands. At twilight, the nomads paused to set up their yurts. Known to the Mongols as *gers*, the term *yurt* was a Turkic word, adopted by the Russians to describe Mongolia's traditional dwellings used since the days of Genghis Khan. Round in shape, they are made of a collapsible framework of wooden lattice covered with

thick felt (a natural conductor of heat and cool) and heavy white canvas lashed with ropes. *Gers* or yurts are remarkably versatile. Varying
from about twenty to forty feet in diameter, they can be erected or taken
down in thirty to forty-five minutes and easily transported by camelback. Heated in frigid weather by an *argul*-burning iron stove or an
open brazier in the center of the room, and ventilated during the summer by raising flaps on the sides, they are perfectly adapted to the country's climatic extremes.

For centuries, the vast majority of Mongolia's inhabitants were nomadic herdsmen. The largest of the country's tribes, the Khalkas, ac-
counted for about 60 percent of the population. But other ethnic groups
were also integrated into an age-old mosaic of cultures—peoples such
as the Kazaks, Durbets, Tuvins, Buriats, Torgots, Zakchins, Charhars,
Uzbeks, and Uryankhas, each with their own dress, customs, and di-
alects, although Mongolia's official language was the Altaic tongue spoken by the Khalkas.

Without industry or extensive agriculture, the nomads depended
for survival on sheep, goats, camels, and horses, along with yaks and a
few cattle that could only be raised in the northern highlands, where
grass was plentiful. Animal products supplied virtually every basic need,
from leather, sheepskin clothing and sleeping bags, and *argul*, or dried
dung, for burning, to goat butter and cheese, the intoxicating kumiss or
airaq, mutton, and sheep's fat, which they particularly relished. Other
necessities such as tea, rice, cloth, tobacco, gunpowder, and metal tools
were acquired from Chinese and Russian traders in exchange for furs,
wool, camel's hair, and horses.

Nomads were generally good-humored, fiercely proud, and hospitable. When visiting an encampment, one was invariably offered
snuff, kept in small, ornate bottles, or food, the alcoholic drink kumiss,
or a mixture of strong tea with milk and laced with rancid goat butter.
On special occasions, a type of "sausage" made from boiled sheep's
intestines stuffed with an unsavory assortment of internal organs was
often served, along with eyeballs of sheep or goats.

Mongols were addicted to athletic contests involving wrestling,
marksmanship, archery, and camel and horse racing—sports deeply
rooted in their legacy as warriors. Since the days of Genghis Khan, their
prowess as equestrians had been legendary. "A Mongol's real home is

the back of a pony," Andrews observed. "He is uncomfortable on the ground. His great boots are not adapted for walking and he is so seldom on foot that to walk a mile is punishment."

In spite of Andrews' admiration for the nomads' independence, physical skills, and likable nature, he was appalled by their uncleanliness: "When a meal has been eaten the wooden bowl is licked clean with the tongue; it is seldom washed. Every man and woman carried through life the bodily dirt which had accumulated since childhood, unless it is removed by some accident or by the years of wear. One can be certain that it will never be washed off by design or water. . . . When Mongols are *en masse*, the odor of mutton and unwashed humanity is well-nigh overpowering."

Venereal disease, especially syphilis, was epidemic; adultery openly practiced; and polyandry was not uncommon. "They are *unmoral* rather than *immoral*," Andrews concluded, "they live like untaught children of nature, and modesty, as we conceive it, does not enter into their scheme of life . . . chastity is not a virtue."

Like sex, death was generally impersonal. Evil spirits were believed to take possession of a corpse, and one was therefore loath to be present even when a relative died. Usually, anyone near death was simply abandoned. If a person died near a village or town, the body was often eaten by wolves, vultures, or huge mastifflike dogs or mongrels, whose taste for human flesh made them extremely vicious. Frequently, the bodies of nomads were placed on carts, which were then driven rapidly over rough terrain in isolated areas. At some point, the corpse bounced off, and fearing that evil spirits would follow them, the drivers never looked back to see where the body landed. Only members of the nobility and the highest-ranking lamas could afford the luxury of a tomb; and near monasteries one saw large numbers of remains eaten by animals—"a curious fate," Andrews remarked, "for holy men who had devoted their lives to the contemplation of eternal mysteries."

On June 13, Merin appeared with news that the caravan was waiting at Mount Arishan with Badmajapoff, who had been spending ten days basking in the hot springs. Some of the camels had again grown weak from lack of food, although for days, they had been traveling through the luxuriant grasslands relished by the nomads' sheep, horses, goats, and yaks. Not so with camels; they refused to eat anything except an unlikely diet of sage, tamarisk, and thornbushes.

Merin also brought unsettling word that the caravan had experienced its first clash with brigands. Several bands of would-be robbers had attacked the procession, only to be driven off by the caravanners' rifle fire. A lone soldier, assigned at the last minute by Mongolian authorities to guard the caravan, boasted that he had hit four or five of the brigands, who appeared to slump in their saddles as they rode away. But the scientists took little comfort in the soldier's bravado; almost certainly, they felt, bandits would continue to stalk the caravan with its salable cargo of gasoline, oil, and tires.

Everyone's outlook improved when Badmajapoff related something he had learned from nomads while at Mount Arishan. Eighty or ninety miles south, according to his informants, lay a region containing bones "as large as a man's body." Having seen no trace of fossil mammals or dinosaurs since leaving Iren Dabasu, "we thought it was merely native exaggeration," Andrews asserted, ". . . but discovered later that they were not so far wrong, after all." Sending the caravan toward the area indicated by Badmajapoff, the rest of the party abandoned the verdant hills above Sain Noin Khan and entered a godforsaken wasteland called the Gordia Basin.

At this point, the expedition had at last penetrated the Gobi. Ahead sprawled an unnerving emptiness. "Desolate it is," Andrews entered in his journal, "but undeniably beautiful & intensely interesting as deserts almost always are. Buttes, mountains, ravines & plains are painted in colors which change every hour of the day—even the sunsets which flood the desert with lights born, it seems, from some distant volcanic crater, and unnatural in their intensity. It is terrible, but fascinating—terrible in its desolation, in its scorching sun & hopeless drought. There is a menace, yet an uncanny excitement, in the mirages of cool lakes with reedy shores, of wooded islets & somber forests. It is like the Lorelei luring men farther & farther into the barren wastes by broken promises."

The expedition arrived in the Gordia Basin in the midst of a violent storm, typical of the Gobi's irascible climate, that struck on June 20, dropping the temperature to freezing and blasting the desert with gusts of hail and snow. When it cleared, Colgate and Badmajapoff set out in search of a guide, returning with an impoverished nomad whose worldly possessions, Andrews reported, included "one wife, one horse, one sheep and one goat . . . [and] like 'Gunga Din,' his clothes consisted of nothing much before and rather less than half of that behind." Yet the tattered nomad was obviously familiar with this intimidating land and, seemingly unafraid of automobiles, he climbed into one of the Fulton trucks, puffing on a cigar given to him by Granger and beaming, as Andrews put it, with "an expression of the most sublime delight." Leading the explorers past a saline lake where nomads were loading bags of salt onto camels, their guide took them to a ridge above a jumble of gray-green ravines and gullies. On the horizon were reddish dunes formed by wind-driven sand piled up in ever-shifting waves. Beyond the dunes rose the jagged profiles of Baga Bogdo and Mount Uskuk, two towering peaks of the Altai Mountains.

While they were camped at a well called Ondai Sair, the Gobi began to reveal its scientific treasures on a scale never imagined by Andrews and his companions. By evening of the second day at Ondai Sair, Granger, Colgate, Larsen, and their energetic Mongol guide had gathered several sacks filled with fossil bones, plus an important collection of ancient fish, crustaceans, plants, and insects preserved in paper-thin shales, including a mosquito and delicate wings of giant mayflies. In the meantime, Berkey and Morris had found the outline of a massive granite batholith (a formation of igneous rock generated by volcanic upheavals but which usually remains below the surface), portions of which they outlined on topographical maps. Andrews pronounced the area a "zoological paradise." His traps snared animals not encountered anywhere else in the desert, among them a rare species of jerboa, a rodent with comically long hind legs that was subsequently named *Stylodipus andrewsi*. After surveying the richness of the discoveries found everywhere, it was soon apparent to the members of the Third Asiatic Expedition that they had stumbled upon the scientific Xanadu for which they had risked their reputations and traveled to the ends of the earth to seek.

A week later, the operation moved to what was called "Wild Ass Camp" because of the number of fleet-footed wild asses in the region. The countryside abounded with rodents, birds, antelope, gazelles, wolves, foxes, and hedgehogs. Even a few lizards scurried among the rocks. Above all, however, the plains thundered with herds of wild asses—fawn-colored with a dark brown stripe down their backs and about the size of zebras. Native only to central Asia and Africa, the Asian variety was almost unknown scientifically.

Here Andrews had an unparalleled opportunity to take notes on their anatomy and behavior, and with Shackelford's camera mounted precariously on a wildly bouncing truck, the two men drove across the gravel plains shooting unsteady footage of the fleeing asses. Andrews clocked them at speeds of forty-five miles an hour; and as a test of one stallion's endurance, he chased it in a Dodge for twenty-nine miles before it dropped from exhaustion. Altogether Andrews and Shackelford killed over twenty stallions, mares, and colts for the Museum's collection.

Granger was kept equally busy at Wild Ass Camp by a succession of paleontological discoveries: the bones of a rare spadefoot toad encased in Oligocene-epoch red clay (approximately 40 million to 50 million years old), and numerous skulls of rodents and other small mammals. In a nearby expanse of badlands, Granger uncovered well-preserved bones of Cretaceous dinosaurs, including a creature with a parrotlike beak, later classified by Osborn as *Psittacosaurus mongolinese*.

On the second day at Wild Ass Camp, Shackelford picked up a large foot bone that appeared to have come from a species of rhinoceros. Not long afterward, Andrews found bits of crumbling teeth close to the site of Shackelford's discovery. Granger spent the better part of four days uncovering the extremely fragile teeth, together with sections of a jawbone in which they were embedded. "No one but a master of the technique like Walter Granger could have accomplished it at all," Andrews commented. Using long-established methods for excavating paleontological specimens, Granger first removed the sand from around the fossils with delicate instruments, often dental or surgical tools, and a camel's hair brush. He then wet each tooth and bone with shellac or a solution of gum arabic and stippled Japanese rice paper into the crevices to bond the surface and protect it from exposure to air. When a sizable portion of the specimen had been coated, it was bandaged with strips of burlap soaked in flour paste. Once this outer "cast" had hardened, it was safe to remove the fossil for transport to the laboratory, where it could be opened and studied.

Andrews readily acknowledged that he lacked the patience for this sort of tedious work. Although he was enthralled with searching for fossils, the time-consuming procedures required to extricate them were incompatible with his restless temperament. "I was inclined to employ [a] pickax," he confessed, "where Granger would have used a camel's hair brush and pointed instruments not much larger than needles." Andrews quoted Granger in *Under a Lucky Star* to the effect that "whenever one of the men is engaged upon the delicate operation of removing a specimen, the chief paleontologist issued an ultimatum to the leader of the Expedition: 'Thou shalt not approach this sacred spot unless thy pick is left behind.'" For years, it was axiomatic in the American Museum's paleontology department that any time an improperly collected or damaged fossil reached the laboratory, it was said to have been "RCA'd"—an allusion to Andrews' heavy-handed approach to collecting, though in time he learned to temper his enthusiasm under Granger's watchful eye.

Eventually, the base of operations was moved to the shores of a glistening body of water twenty miles to the southwest. Roughly three miles long by two miles wide, it was known as Tsagan Nor (White Lake), and it teemed with flocks of waterbirds and small fish. On July 3, Andrews had sent Colgate, Larsen, and Badmajapoff with one Dodge automobile back to Urga. Larsen and Badmajapoff were leaving the expedition, forced to resume their business interests, which they had neglected for over six weeks. Andrews had grave misgivings about sending a single automobile on the eight-hundred-mile-drive, especially in view of continual reports of bandit raids, but Colgate was confident he could get through safely by following the route already blazed by the expedition on its southward journey. As a matter of fact, he returned in only twelve days, bringing a leather pouch bulging with mail, including a letter from Yvette in which she reported that she was safely back in Peking after an uneventful trip from Tuerin.

Andrews declared the Tsagan Nor basin "exceptionally well suited to our purposes." Almost without a break, the scientists were kept busy examining its rich fauna, including massive herds of wild asses and gazelles, uncharted geological deposits, and a variety of fossils. On July 28, the keen-eyed Shackelford, fast becoming a stellar fossil hunter, made what later proved to be a startling discovery. In a dry streambed, he came upon the humerus of an enormous unidentified mammal. Within several days, Granger, Shackelford, and their Chinese chauffeur Wang returned from a prospecting trip laden with fossils that included the end of a humerus, one side of a lower jaw, and other skeletal parts. Actually, it was Wang who first spotted the bones and alerted Granger to their location. All of them were similar in size to the humerus found earlier by Shackelford near Wild Ass Camp, and the jaw matched the one Andrews had discovered in the same locality, which Granger had so meticulously excavated.

Anxious to collect more of these gigantic bones, Shackelford, Wang, and Andrews returned to the Loh Formation near Wild Ass Camp the next day. Hardly had Andrews arrived before he was brushing away sand from a colossal skull. Summoning Granger to the site, he labored for four days exposing the spectacular fossil; eventually, it was sealed against exposure, encased in flour-soaked burlap, and loaded into two boxes for shipment to New York.

Granger and Osborn surmised that this Oligocene-aged beast was a rhinoceros-like creature belonging to a group of ungulates—animals related to horses, tapirs, and camels. Initial studies established that the massive skeletal remains were indeed those of an aberrant rhinoceros, which Osborn named *Baluchitherium mongolinese*. Some years earlier, in 1911, this animal was first reported by C. Forster Cooper of Cambridge University. While exploring in the Bugti Hills of eastern Baluchistan, he had stumbled upon bones of two closely related if not identical species of rhinoceroses—one designated by Cooper as *Baluchitherium osborni* or the "Beast of Baluchistan," and another known as *Paraceratherium*. Then, in 1915, a Russian paleontologist, A. Borissiak, discovered teeth and skeletal parts of a creature he called *Indricotherium* near Turgai in Russian Turkestan. Borissiak's specimen was virtually indistinguishable from *Baluchitherium* and *Paraceratherium*, and although scientists still find grounds for debate, all three of these creatures are now generally referred to collectively as *Indricotherium*. Andrews, Granger, and Osborn never abandoned the term *Baluchitherium*.

Long touted as the largest land mammal ever to roam the earth, early estimates based on incomplete skeletons suggested *Indricotherium* may have stood seventeen feet at the shoulders, attained a body length of twenty-four feet, and weighed approximately thirty tons. (These dimensions have now been somewhat reduced, though it undoubtedly attained an enormous size.) Its skull measured around five feet in length, and its prehensile lips and long neck allowed it to browse in trees like present-day giraffes.

The expedition abandoned its camp at Tsagan Nor on August 13. Impelled by "an autumn sharpness in the air," and a thousand miles of virtually unknown country between Tsagan Nor and Kalgan, Andrews was anxious to proceed eastward. The caravan departed for a rendezvous point near the base of the eight-thousand-foot-high Artsa Bogdo, another of the Altai's sculpted ranges. Taking enough gasoline for several weeks, the cars set out along the Kweihwating-Kobdo trail, which skirted the eastern end of Baga Bogdo and passed within fifty miles of Artsa Bogdo. Eventually, the explorers planned to intersect the well-traveled Sair Usu trail and follow it back to Kalgan.

From the start, the journey proved scientifically rewarding. Engaging a Mongol guide, Andrews headed off to hunt ibex and bighorn sheep on Artsa Bogdo's rocky slopes, killing numerous superb specimens. Meanwhile, Berkey and Morris traversed the desert south of Artsa Bogdo and explored a maze of towering peaks and escarpments, the Gurban Saikhan (or Gurvan Saichan), one of the most rugged of the eastern Altai's ranges. In addition to performing geological studies, they collected late Cretaceous fossils and a piece of red limestone containing Paleozoic invertebrates—crinoids and corals.

While these activities were under way, Granger had gone northward to reconnoiter an area later named the Oshih Basin, where he reported finding a dinosaur almost literally "under every bush." Scattered over this inferno-like landscape—gashed with ravines, canyons, and gullies from which rose a red sandstone mesa capped with black lava— were hundreds of fossilized teeth and bone fragments. Granger excitedly displayed two skeletons of the small dinosaur, *Psittacosaurus*, with its whiplike tail and a beak similar to a parrot, one of which was in nearly perfect condition. Upon closer inspection the following year, the Oshih Basin would sorely disappoint Granger. On the surface, at least, it yielded nothing of interest and its fossils were too fragmentary to be of value.

Increasingly, the approach of winter left no choice but to abandon further explorations. "The persistent chill," Andrews wrote, "was a warning to put the long stretch of unknown desert behind us without delay, and the appearance of a blanket of new snow on Baga Bogdo, eighty miles westward, forced me to announce that nothing except the discovery of the 'Missing Link' himself would keep us after September 1."

But, fortuitously, Andrews' plans did not go as anticipated. Losing their way in a nearly featureless expanse of desert, the scientists spent three days in a futile search for a route northward to the Sair Usu trail. Andrews had expected that nomads would be able to give them directions, but they traveled nearly a hundred miles before a cluster of yurts was finally sighted. Leaving Shackelford to await the other cars, Andrews drove over to confer with the Mongols.

Moments later, Shackelford wandered away to inspect an unusual outcropping. After walking a few yards, he found himself looking into a sweeping basin filled with spectacular formations cut by erosion into massive walls of reddish-orange sandstone. "Almost as if led by an invisible hand," Andrews recounted, "[Shackelford] walked straight to a

Psittacosaurus

First unearthed by Walter Granger in the Gobi's Oshih Basin in 1922, Psittacosaurus belonged to a group of dinosaurs classified as neoceratopsians, a characteristic of which was a bone on the front of the skull that resembled a parrot's beak.

small pinnacle of rock on top of which rested a white fossil bone. Below it the soft sandstone had weathered away, leaving it balanced and ready to be plucked off." Granger identified Shackelford's well-preserved trophy as a reptilian skull, although one unlike any he had ever seen. After pitching their tents at the basin's edge, everyone spent the remaining daylight hours scouring the region. The ground was littered with bones; others could be seen eroding out of the formations. Granger even picked up part of a fossilized eggshell, which he assumed had been laid by some long-extinct bird. In short, Shackelford's accidental discovery was a fossil hunter's utopia!

"[It] is one of the most picturesque spots that I have ever seen," extolled Andrews. "From our tents, we looked down into a vast pink basin, studded with giant buttes like strange beasts, carved from sandstone. . . . There appear to be medieval castles with spires and turrets, brick-red in the evening light, colossal gateways, walls and ramparts. Caverns run deep into the rock and a labyrinth of ravines and gorges studded with fossil bones make it a paradise for paleontologists." Originally called Shabarakh Usu (Place of the Muddy Waters) because of a spring-fed lake in an adjacent valley, this wondrous place, so euphorically described by Andrews, is known today as Byan-Dzak after the scrubby zaksaul trees that flourish in nearby sand dunes. But Andrews, Granger, and their fellow scientists, awestruck by the basin's brilliant red and orange formations, would christen it the "Flaming Cliffs," the name by which it is still known to every paleontologist. "We could hardly suspect that we should later consider it the most important deposit in Asia," wrote Andrews, "if not in the entire world."

At a remote lamasery called Ongin Gol-in Sumu, some distance northeast of the Flaming Cliffs, the lamas directed the explorers to the Sair Usu trail and reported that Merin and the caravan had passed that way two days earlier. Forty-seven miles farther on the men sighted half a dozen ruined mud huts and two abandoned temples in a sandy basin next to two wells. "Such was Sair Usu," reported Andrews, once the confluence of major caravan routes from central Asian cities to China. "A little to one side of the nearest well stood the blue tents of our caravan."

Instructing Merin to proceed straight to Kalgan, Andrews unloaded gasoline and food for the 536-mile drive before sending the caravan on its way. Regardless of Andrews' anxiety regarding the onset of winter,

Granger, Berkey, and Morris—none of whom was experienced enough with Mongolia's dangerous climate to appreciate the risk—insisted on stopping every few miles to inspect promising formations. Scarcely eleven miles from Sair Usu, Granger dashed off to explore a cliff of fossiliferous limestone studded with Paleozoic invertebrates. Farther along the trail, a line of ridges jutted out of the desert, marked by red, yellow, and gray striations. Berkey, Granger, and Morris prospected the red sediments at its base without success. Again, it was Shackelford, examining the overlying yellow and gray strata "with the instinct of a pointer dog for game," observed Andrews, who quickly filled his pockets with bones and teeth. Andrews found pieces of fossilized turtle shell, and when a fine specimen of a rhinoceros, along with other fossils, was uncovered, it was decided to camp for three days at the bluff, which was called Ardyn Obo because of an impressive *obo* or religious shrine on its summit, surrounded by many smaller ones.

At last, the explorers could delay no longer. Icy winds and snow flurries forced them to hastily abandon Mongolia for China. Late in the afternoon of September 18, the expedition drove triumphantly through Kalgan's gateway, exactly five months after the venture began. Crowds of Chinese lined the streets to greet the cars, with their Klaxon horns blaring as they roared into the Anderson, Meyer and Company compound, where a grinning F. A. Larsen was anxiously awaiting their arrival.

After a bath and change of clothing, everyone gathered at the British-American Tobacco Company's headquarters, where a dinner had been arranged. "Our efforts to meet the requirements of civilization were almost pathetic," wrote Andrews. "Each of us had some article of adornment which he had been cherishing for the home-coming. Shackelford appeared in a wonderful blue shirt. I had a purple necktie, while Granger and Colgate each produced a new pair of shoes. When we came into the dining room for tea, where half a dozen visitors had assembled to welcome us, we all felt decidedly uncomfortable."

It had been impossible to advise the outside world regarding the expedition's activities by mail or cable during its five months in the field. And it would be some weeks yet before the specimens arrived in New York, where they could be examined in the Museum's laboratories. Knowing that Osborn, his colleagues, and the public at large were anxiously awaiting news regarding the outcome of the explorers' venture, it was decided to release a cable outlining a preliminary summary of their accomplishments.

The day after returning to Peking, Andrews called the staff together in the drawing room of the headquarters compound. "We were dressed and shaved," he recounted in *Under a Lucky Star*, "and had taken on the restraints of city life along with its habiliments. The men sat on the silk sofas beside the fire and looked at each other almost as though they were strangers. No one said anything." Realizing that something had to be done, Andrews instructed his Number One Boy, Lo, to mix a round of cocktails—"and make 'em strong," he said, even though it was only eleven in the morning. "We toasted each other and the success of the expedition. . . . Lo, unobtrusively, saw that every glass was filled as soon as it was emptied and in fifteen minutes the tension had broken."

For two hours, they worked at composing a suitable cable. Andrews and the Museum had agreed from the outset that political intrigue in China and Mongolia made it advisable to maintain secrecy by transmitting all messages in code, using the cable address "Museology," Peking, and codes compiled from a book known as *Bentley's A. B. C. 5th and 6th editions*. But for this cable, which was meant as a press release, no code was deemed necessary. "We had made scientific history," Andrews announced, "and I wanted to get it into the public record as soon as possible. . . . When we got through, the cable was something of a masterpiece. It told everything I was sure the newspapers would want to know—or at least ought to want to know."

Sent out immediately to the Museum, *Asia* magazine, and the press wire services, Andrews made certain that the cable's underlying message could not be overlooked: those skeptics who had asserted that searching for fossils in the Gobi's wasteland was sheer folly were dead wrong!

The cable enumerated an astonishing array of accomplishments. There was, to be sure, one conspicuous omission: the absence of any traces of the much-ballyhooed "Missing Link." But this disappointment was temporarily obscured by the explorers' other feats as set forth in Andrews' paraphrase of the cable:

> Scientific results had surpassed our greatest hopes, yet we
> knew that we had only scratched the surface. The expedition

had traveled . . . to the Altai Mountains, where vast fields rich in Cretaceous and Tertiary fossils had been discovered, including a fine skull of and parts of the skeleton of *Baluchitherium*, the largest known land mammal.

We had also obtained complete skeletons of small dinosaurs, and parts of fifty foot dinosaurs, skulls of rhinoceros, hundreds of specimens, including skulls, jaws, and fragments, of mastodon, rodents, carnivores, horses, insectivores, deer, giant ostrich, and egg fragments. We had found wonderfully preserved Cretaceous mosquitoes, butterflies, and fish, unknown reptiles, titanotheroids, and other mammals.

The geologists had identified extensive deposits of Devonian, Carboniferous and Permian age Palaeozoic rocks. They had measured twenty thousand feet across upturned edges of Jurassic strata. An enormous granite bathylith had been discovered comparable to the Laurentian bathylith of Canada, with a wonderful development of roof pendants and contact metamorphism. We had the longest detailed topographic route map and continuous geologic section ever made on reconnaissance. We had mapped a thousand-square-mile strip in the type region of Mongolian geology.

Our photographer had obtained twenty thousand feet of film of all details of the work of the expedition, and feature films of every phase of Mongol life. He had filmed large herds of antelope and wild asses, never before photographed alive. He had obtained five hundred still photographs.

The zoological work had been extraordinarily successful. We had secured the largest single collection of mammals ever taken from Central Asia, including many new species, and material for fine habitat groups of the wild ass, antelope, ibex, and mountain sheep for the Museum's Hall of Asiatic Life.

Clifford Pope working in North China and the Ordos desert had obtained a splendid collection of fish and reptiles. He had made a comprehensive zoological survey of Shansi Province.

Following the cable's release, the explorers were deluged with requests for interviews. An ecstatic Osborn responded with characteristic élan: YOU HAVE WRITTEN A NEW CHAPTER IN THE HISTORY OF LIFE UPON THE EARTH. Congratulatory messages poured in from scientists and geographical societies in England, America, France, Germany, Australia, Hungary, and Sweden. "I must say," Andrews could state with satisfaction, "that those scientists who had béen loudest in their prediction of failure were the first in admitting that they had been wrong." It also became evident that the expedition was inadvertently tinged with commercial implications, something Andrews had feared from the start. "All unwittingly," he noted, "we had opened Mongolia for motor transportation. Representatives of Chinese importing firms asked how they could get cars to various points in the Gobi to bring out valuable furs, contract for hides, camels, and sheep's wool and ponies; what routes to take, where to send gasoline and a dozen other questions."

For Andrews, the undertaking was a dazzling personal triumph. His risk-laden plan to drive automobiles across the Gobi was a brilliant success. His seemingly ill-considered concept of resupplying the expedition with a camel caravan and using it to carry specimens back to China went off without a serious flaw, as did the innovative idea of dividing his field parties into independent units. Equally important, he had justified the blind-faith financial backing received from Morgan, Baker, Dodge, Rockefeller, James, Davidson, Frick, and the legion of other supporters, who risked scathing ridicule had the project failed. Instead, the Third Asiatic Expedition was on everyone's lips, and Roy Chapman Andrews was the man of the hour. An incredulous public was captivated by his audacity, ingenuity, courage, genius for organization, and splendid leadership in the face of the Gobi's political obstacles, bandits, and physical rigors. In short, Andrews' already adventurous persona—so enamored of the exotic, the dangerous, and the unknown—exploded overnight into full-fledged celebrity. America's newest hero would soon realize that the relentless glare of fame could become irresistibly addictive.

CHAPTER 12

Soon after returning to Peking, Berkey left for New York to resume his geological work at Columbia. Frederick Morris stayed behind to complete the research he and Berkey had begun in Mongolia the previous summer. Meanwhile, Granger, again accompanied by his English-speaking guide James Wong and most of his assistants from the winter of 1922, set out on the Yangtze River bound for Szechwan Province to continue studying the fossil pits at Wanhsien and Yen-ching-kou. This time, Granger took along his wife, Anna, who not only observed the digging, but wrote several articles on her adventures that appeared in *Natural History*.

As expected, the political machinations in the region had remained unchanged from the previous year. Wu Pei-fu's soldiers were continuing their half-hearted war against the local militia. Brigands were terrorizing the countryside with reckless abandon, repeatedly firing at the junk that carried the scientists up and down the Yangtze. Still, despite these risks, there were no casualties and the party returned to Peking in March with numerous crates of Cenozoic mammal bones—badly broken after being hacked out of the pits with picks and hauled up in baskets by the Chinese "dragon bone men."

With the stunning success of the first expedition behind him—and basking in an outpouring of accolades—Andrews and Yvette reveled in Peking's worldly pleasures. Their drawing room in the headquarters compound on Bowstring Street became a magnet for prominent visitors to the city. Together with a group of friends, they rented a secluded temple in the thickly wooded Western Hills to use for outings and picnics, a luxury widely in vogue among the foreign colony. Andrews later rented a larger and more comfortable temple at a cost of $240 a year. Known as the "Temple of the High Spirited Insects," its name so enchanted Andrews that he called his polo team the "Peking Insects." As always, there were lavish dinner parties in private homes, legation receptions, dancing at Peking's hotels, and sporting events that, for Andrews, revolved around his obsession with horses: training, racing, jumping, and point-to-point hunting. Above all, polo had become "a passion with me," he wrote. Excellent ponies were cheap (about $100 on average), "[and] by dint of hard practice I got to be pretty good and always rode as one of the four which represented Peking in the Interport matches against Shanghai, Tientsin, Hankow, and regimental teams from all over the east."

On a frigid December night in 1922, Andrews and Yvette glimpsed a curious footnote to the demise of the Manchu Dynasty. They attended the only public festivity surrounding the marriage of the last emperor, Pu Yi, to Wan Jung (Beautiful Countenance), the sixteen-year-old daughter of a wealthy Manchurian family from Tientsin. A sensual beauty, Wan Jung, nicknamed Elizabeth by the emperor, had been well educated in an American missionary school and by private tutors. She spoke fluent English, and was considerably more worldly than the reclusive emperor, who, after all, had rarely left the confines of the Forbidden City. Under a bright moon, with the temperature below freezing, the princess-to-be was paraded through Peking's streets in a gilded sedan chair embellished with silver birds, symbols of good fortune. Reaching a gateway to the Forbidden City, she was carried through the arched opening by eunuchs, who escorted Wan Jung, outfitted in opulent silk robes and jewels, to the Palace of Earthly Pleasures. Here the emperor perched on the venerable Dragon Throne in lavish ceremonial regalia and, surrounded by nobles and family, awaited his bride's arrival.

In an adjacent hall, a group of foreign dignitaries had gathered by special invitation. Andrews and Yvette were accompanied by the American minister, J. V. A. MacMurray. Everyone milled about shivering in the bitter cold and sipping tea until a bugle signaled the approach of Wan Jung's sedan chair on her way to the Palace of Earthly Pleasures. The guests quietly lined up along a marble causeway over which only the emperor normally passed. "There," wrote Andrews, "[the eight bearers] paused, then moved swiftly through a lacquered doorway. . . . As one man, the Manchus dropped upon their knees, foreheads touching the ground. Automatically I followed suit. Only the ministers remained partially erect, bent over in deep bows. . . . It was all over in a few seconds. The chair passed between the kneeling throng, entered a yellow carved gate into an inner court and was lost to sight."

No foreigners were present at the actual marriage ceremony, enacted in closely guarded privacy. But a chosen few, including Andrews and Yvette, were admitted to an afternoon reception. "The Emperor," Andrews remembered vividly, "his weak eyes hidden behind dark spectacles, stood beside his bride, impassive, unsmiling. She was lovely. A little heart-shaped face, soft brown eyes, and a body slender as a willow wand. Her hair was done in the traditional style of a bride whose marriage has been consummated. The emperor spoke English fairly well . . . but he was shy of speech. I offered my congratulations in Mandarin Chinese. He answered in the same language."

Years later, in 1932, Andrews arranged an audience with Pu Yi in Manchuria in a futile attempt to enlist his aid in regaining entrance to the Gobi, which had been sealed off by antiforeign forces. By then the Japanese had lured the emperor, who was totally devoid of power, to Manchuria with false promises of restoring the monarchy, part of a scheme to control northern China and eventually conquer the entire country. Pu Yi, his life in shambles, proved utterly helpless. Andrews found him a "pathetic puppet, surrounded by hissing Japanese; an ineffectual boy caught up in the maelstrom of a political flood which whirled him around like a straw upon the waters."

In the spring of 1923, preparations began for the second journey to the Gobi. The main order of business obviously entailed a thorough investigation of the most promising fossil beds discovered during the earlier reconnaissance. Accomplishing this meant that certain changes in the expedition's scientific makeup were necessary. Most important, Granger needed several more experienced collectors. A cable was dispatched to Osborn requesting three of the Museum's ablest men: two Danish-born paleontologists, Peter Kaisen and George Olsen, along with a skilled collector named Albert Johnson, a rather rough-hewn character who had worked extensively with the legendary paleontologist Barnum Brown. And since the upcoming journey would largely retrace territory already photographed, it was decided not to include J. B. Shackelford this time around.

After the experience of the first season, Andrews concluded that two mechanics were needed to effectively maintain the automobiles. As Bayard Colgate had set out to complete a round-the-world trip he was planning before the 1922 season, Andrews engaged J. McKenzie Young, a member of a Marine detachment in Peking, who was to serve as chief of motor transport from 1923 to 1930. He was assisted in 1923 by an expert mechanic, C. Vance Johnson, recruited from an automobile distributor in Peking. "Mac," as Young was known, was a soldier of fortune, "handsome as an Adonis," Andrews remarked, and bursting with irrepressible charm. He became one of Andrews' closest friends and the subject of a glowing profile in his book *This Business of Exploring* (1935).

The son of a Presbyterian clergyman, Mac Young was attending school in Canada when World War I broke out. He obtained a commission in the Canadian army, expecting to be sent to Europe. When his request was denied, Mac deserted and joined a unit bound for France. As punishment, he was court-martialed, reduced to the rank of private, and ordered to the front, where shrapnel tore through his leg a day and a half later. Returning to the trenches after a lengthy recuperation, he was immediately wounded again. Next, he spent a year in France manning howitzers, survived a direct hit that killed every other man in his battery, and was promoted to top sergeant. When the war ended, he had just begun training with the United States flying corps. Back in America, he worked briefly in a bank before enlisting in the Royal Canadian Mounted Police. With two friends, he then tried his luck at trapping, but lost his furs and equipment when Indians raided their camp. Lured by a sign reading JOIN THE MARINES AND SEE THE WORLD, Mac enlisted on the condition that he would be posted to China. He ended up in the Legation Guard in Peking as a corporal in charge of maintaining the unit's cars and trucks because of his expertise as a mechanic. Here he met Andrews and was assigned to the Central Asiatic Expeditions for the next eight years.

Once again, the headquarters began to bustle with activity, and massive stores of provisions and equipment were assembled in the courtyard. The staff was increased to forty men, with the addition of more paleontologists and technicians. Granger's three colleagues, Peter Kaisen, George Olsen, and Albert Johnson, arrived from New York, as did Berkey, who was reunited with Frederick Morris. One of the mess boys, Kan Chuen-pao, whom everyone called "Buckshot," had demonstrated such aptitude in handling fossils during the winter's excursion to the pits at Yen-ching-kou that Granger decided to make him a full-time field assistant. Another young Chinese, Liu Hsi-ku, originally hired as a mechanical assistant, also developed into a skilled fossil collector, and in 1924, both men were sent to the Museum in New York for special training.

On April 2, the supply caravan departed from Kalgan, again led by the indomitable Merin. Fifteen days later, on the seventeenth, the staff left Peking, touching off a deluge of publicity in newspapers in the United States, Europe, and China. For Andrews and Yvette, it was another difficult parting. While five months of high adventure lay ahead for Andrews, Yvette had no choice but to suppress her loneliness and apprehension over her husband's safety by running a sprawling, servant-filled house, caring for George, and maintaining a busy social life. It was increasingly clear that the expeditions were inexorably taking a toll on their marriage.

The explorers were delayed in leaving Kalgan for two days by a heavy snowfall that had turned the road into Mongolia to a "mass of gluelike mud." And an ever-present hazard, bandits, was plaguing the region. A week earlier, two Russian cars loaded with furs had been ambushed by brigands dressed as Chinese soldiers. Everything—furs, money, and personal possessions—was stolen and the traders were stripped of their clothing and left naked; one man was brutally slaughtered. Andrews was further unnerved by the death the previous winter of his close friend Charles Coltman, who was inexplicably murdered by Chinese soldiers. All the robberies were occurring in an area just north of Kalgan, which was under the ineffectual control of the *tuchun* of Chahar. Every few weeks, freight cars loaded with "brigands" were brought with much fanfare to Kalgan by the *tuchun's* soldiers, many of whom were bandits themselves, taken to the dry riverbed that ran through the center of the city, and shot. But these executions failed to stop or even impede the raids.

Finally, on April 19, still traveling under the permits or *huachos* issued a year before in China, the expedition entered the field, passing through the Great Wall at Wanchuan. Following their earlier route from the inn at Miao Tan to P'ang Kiang, the procession headed directly to Iren Dabasu, where the caravan was to have awaited their arrival. When it was nowhere in the vicinity, Andrews became gravely alarmed—and with ample justification. The area was so dangerous that a Chinese military garrison at Miao Tan was ordered to patrol the road ahead of the cars, although there was no way to protect the caravan. At one point along the way, about twenty brigands stealthily appeared among the hilltops, ". . . but our cars," reported Andrews, "bristling with rifles, did not impress them favorably and they left us severely alone."

After waiting another week, Granger and Morris drove seventy miles in the direction the caravan was supposedly traveling, but did not locate it. Andrews feared it had been captured or driven into the desert, thereby ending the expedition before it ever started. "I made up my mind that if it was not heard from soon I would take three or four of our men who were simply spoiling for a fight, follow the caravan on horseback from the point where Merin had last been seen, and recapture it."

But the wily Merin had escaped unharmed. One evening, a car rolled into camp bringing news that the caravan was twenty miles away. When he learned brigands were watching the trail ahead of him, Merin reported that he had slipped off into the desert. From then on, he traveled only at night from well to well, camping during the day in sheltered hollows where the caravan could not be seen easily.

With the caravan safely in tow, another obstacle arose when Larsen sent word that the Mongolian government—contrary to the agreement signed the previous year—was threatening to withhold the explorers' passports. Andrews was predictably furious at this news, which he regarded as an outright betrayal. Even without a passport, he resolved to drive to Urga and personally intercede in the matter. Again, however, Larsen's diplomacy, and the support of a few sympathetic Mongol officials, overcame the Bolsheviks' resistance, though by now they were more entrenched in the country than ever. Andrews considered it "nothing short of a miracle" when the passports arrived by automobile on June 11. Once more, the gateway to the Gobi was opened despite unfavorable odds.

It was unanimously agreed by everyone that the primary objective of the 1923 expedition should be the Flaming Cliffs. Shackelford's unidentified dinosaur skull, the egglike fragment found by Granger, and the prolific array of bleached bones littering the ground and eroding out of the sculptured formations offered an irresistible lure. Moreover, the decision to focus on this region was also prompted by Granger, William K. Gregory, and another of the Museum's paleontologists, C. C. Mook. After studying the reptilian skull found at the Flaming Cliffs by Shackelford in 1922, they concluded that it represented either the direct ancestor or a close relative of the mighty horned dinosaurs known as ceratopsians, the most famous of which, *Triceratops*, had roamed the western United States and Canada during the Cretaceous period. Accordingly, Shackelford's specimen was named *Protoceratops andrewsi*, and the Museum was anxious to secure more examples for study.

But Granger's first objective was to revisit Iren Dabasu for a closer look at the mammal and dinosaur remains he, Berkey, and Morris had stumbled upon there four days into the 1922 expedition. Granger also felt that Iren Dabasu was an ideal place to give Olsen, Kaisen, and Johnson their initial taste of field experience in Mongolia before penetrating the Gobi's interior.

In the meantime, Andrews, the assistant mechanic C. Vance Johnson, and Liu returned to Kalgan in two cars to load supplies and equipment that had arrived too late to go with the caravan. The trip was uneventful until, on their return to Iren Dabasu, with Johnson and Liu traveling about a mile behind him, Andrews caught the glint of sunlight on a rifle barrel. It came from a hilltop not far from where the two Russian cars had been robbed. Soon he could see a single horseman watching his approach through binoculars. Andrews surmised that the man was a brigand waiting to alert his companions hidden among the rocks. Wishing to avoid the fate of the Russian traders, Andrews decided to strike first. Drawing his Colt .38 revolver, he fired twice at the horseman without trying to hit him. The rider suddenly vanished. Moments later, as the car topped the rim of a valley, Andrews spotted three bandits at the base of the slope.

It would have been difficult to turn the car and run without exposing myself to close-range shots [he later wrote], and, knowing that a Mongol pony would never stand against the charge of a motor car, I instantly decided to attack. The cut-out was wide open and, with a smooth down-hill stretch in front of me, the car roared down the slope at forty miles an hour. The expected happened! While the brigands were attempting to un-ship their rifles which were slung on their backs, their horses went into a series of leaps and bounds, madly bucking and rearing with fright, so that the men could hardly stay in their saddles. In a second the situation had changed! The only thing the brigands wanted to do was get away, and they fled in panic. When I last saw them they were breaking all speed records on the other side of the valley.

Aware of the increased threat of bandits, Andrews kept his men on a high state of alertness, especially near caravan routes where commercial goods entering or leaving China provided an endless source of vulnerable loot that could easily be disposed of in Kalgan or Peking. Invariably, brigands were Mongols, or renegade Chinese soldiers disguised as Mongols, though occasionally displaced Russians survived by banditry. Typically, they traveled in bands of about a dozen men; occasionally, several hundred would unite and terrorize whole areas. Even large caravans were ambushed routinely by bandits concealed among hills, rocks, or ravines, who plundered and murdered without mercy. Nor was it uncommon for entire villages to be raided. One therefore quickly became suspicious of any armed horsemen, even harmless nomads, until their identity could be verified. In areas where bandits were known to be operating, Andrews always posted two or three men in shifts to guard the camp at night, and the explorers' impressive array of rifles and pistols were clearly visible to anyone who encountered the expeditions.

During the time Andrews was away in Kalgan, the scientists had been having their share of excitement. Scarcely had they begun to explore new formations around Iren Dabasu when Albert Johnson spotted an unimpressive three-inch piece of bone jutting out of the ground. On a hunch, he began clearing the earth from around it, exposing an extensive quarry of dinosaur skeletons representing a variety of species entangled in a jumbled mass. Both flesh-eaters and herbivores were intermingled in the heterogeneous chaos of fossils. They appeared to have been caught in the swirling action of water, perhaps a fast-flowing stream or eddy in which the animals had become mired. Evidence that the area was once wet and thick with vegetation was clear from the large number of crocodiles, turtles, and iguanodontids—lumbering dinosaurs that fed on leaves and succulent plants that grew in wetlands near lakes, marshes, and rivers.

Kaisen and Olsen found another quarry not far from Johnson's, where dinosaur bones were intermingled like "jackstraws." Sorting them out was a tedious job, and some were so tightly cemented together that they could not be separated in the field. In such cases, large blocks of matrix were removed intact and shipped to the Museum, where laboratory technicians could painstakingly extricate the bones using special instruments.

Leaving Johnson and Kaisen at Iren Dabasu to complete work on the quarries, the expedition moved to a location known as Irdin Manha, or the Valley of the Jewels, named for the countless multicolored quartz pebbles that littered the ground. Here Buckshot, Granger's Chinese assistant, made a puzzling discovery—a beautifully preserved skull, three feet long, of an unidentified mammal. Andrews believed it belonged to a variety of "the great carnivore *Mesonyx*, an odd looking creodont of gigantic size." At first, Granger disagreed. It was, he thought, more akin to an *Entelodon*, a strange piglike creature with hooves. But after studying a detailed drawing of the skull made in situ by Morris, W. D. Matthew wrote from the Museum supporting Andrews' original conclusion: the specimen was that of an early creodont of the family Mesonychidae. Osborn would later name the species *Andrewsarchus mongoliensis*, and judging from the skull's proportions, he declared that it had been greater in size than an Alaskan brown bear, although most of what is known about *Andrewsarchus* is based on this single skull. Irdin Manha also abounded with the bones of strange hoofed beasts called titanotheres, which had once roamed the Western Hemisphere during the Eocene and Oligocene epochs, roughly 50 million to 25 million years ago.

Departing Irdin Manha, the men spent two weeks at Ula Usu— Well of the Mountain Waters—where the skulls of twenty-seven more titanotheres were collected. But the work at Ula Usu was hampered by constant sandstorms. By late afternoon of the second day at the camp, a strong wind suddenly turned into a fierce gale. Yellow clouds of dust became what Andrews described as "a thousand shrieking demons." Groping his way back to camp, he and a few other men lay buried under wet clothing. One by one, everybody straggled into camp, except Granger, who did not appear until the storm subsided several hours later. He was brown from head to foot, caked in dirt from a pit where he lay half-buried trying to protect a titanothere skull. "[Granger and I] looked at each other and burst out laughing. 'Great Gods! Am I as dirty as you are Roy?' he asked. I assured him he was, only more so. When he looked at himself in a mirror he grunted disgustedly, 'That finishes it. The Mongols have the right idea, no more baths for me. What's the use. I'm going to bed.'"

Proceeding to Ardyn Obo, the eroded maze where fossil deposits had been discovered the previous summer, Andrews found the supply camels waiting not far away, severely weakened by the ravages of drought. "It was imperative that the caravan reach the Flaming Cliffs with gasoline and food . . . otherwise, we would be cut off from the most promising fossil field in all Mongolia," he wrote. "I made up loads for sixteen camels, consisting chiefly of gasoline, and just enough food to carry us along on short rations. These were to be put on the strongest camels and pushed through at all costs. The other camels with their [boxes] were to be brought as far as they could go; if they died their loads were to be left. Under these circumstances, Merin felt certain he could at least get the sixteen loads to us. . . .

"We packed enough gasoline in the cars to take us to our destination, and food for a month. It was little enough . . . but even so the motors were greatly overloaded and could not carry another pound." Gas and provisions for the return trip from Ardyn Obo to Kalgan were stored in a Lamaist temple to be picked up on the way out of the desert.

On July 3, the cars began the four-hundred-mile journey to the Flaming Cliffs. The trail was exceedingly rough, with the cars repeatedly sinking up to their running boards in sand. One truck finally had to be unloaded and carried up a long slope so it could be used to pull the others over the treacherous surface. Now and then, Andrews found caravan routes that avoided these obstacles, but to his consternation, the nomads who lived beside these trails often had no idea where they led. Occasionally, some of the natives encountered along the way still panicked at the approach of automobiles. For the most part, however, word of the "dragon hunters'" presence had spread far and wide by means of what Andrews called "well telegraphs"—news carried from one well to another by nomads who gathered at these meeting places to collect water.

By this time, the excruciating eye inflammation that had begun to affect Andrews' vision in 1922 was becoming acute. The desert's shimmering glare, aggravated by dust, had left him unable to work on his zoological notes or the steady flow of articles he was writing for *Asia* and other publications. Even his journal entries had to be periodically abandoned.

The Flaming Cliffs were reached at 3:15 on the afternoon of July 8, 1923. "Everything was exactly as we left it on our last visit," Andrews noted. "The marks of our tents and the motor car tracks were almost as distinct as though they had just been made." On a sun-scorched plain facing the vividly colored formations, the dark blue tents were arrayed in rows, the cars lined up to one side, and piles of boxes brimming with supplies, gasoline, and scientific paraphernalia were unpacked. Everyone then set about searching for fossils. In less than an hour after the party's arrival, one of the most celebrated paleontological expeditions in history was launched. Before the staff returned for dinner that evening, everyone had begun to excavate dinosaur skulls—mostly those of *Protoceratops,* which littered the formations, leaving no question that the region had swarmed with these bizarre creatures during Cretaceous times, some 145 million to 65 million years ago.

As skeletons began to emerge along with skulls, it was evident that *Protoceratops* was relatively small, six to eight feet long on average. The skulls of mature adults measured about twenty-three inches in length, and had a powerful, curved beak resembling that of a parrot. Extending from the top of the skull over the jaws and neck was a thick bony plate. There was no doubt that they had walked on all four feet, though if necessary for defense or feeding they could almost surely have risen up and propelled themselves on their powerful hind limbs.

Five weeks of exploration at the Flaming Cliffs turned up seventy skulls and fourteen skeletons of *Protoceratops* in virtually every stage of their growth cycle. It would, however, prove impossible to confirm

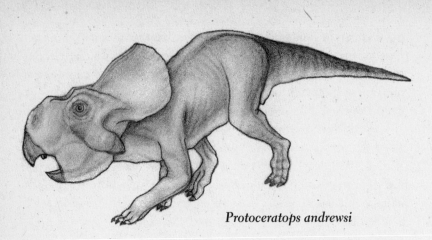

Protoceratops andrewsi

First discovered at the Flaming Cliffs in 1922, this horned Cretaceous dinosaur was initially believed to have been the direct ancestor of the great ceratopsians such as Triceratops that once flourished in North America. It was also thought—mistakenly as it turned out (see Epilogue)—to have laid the famous eggs found in such abundance at the Flaming Cliffs and elsewhere in the Gobi Desert.

whether these were the earliest of the ceratopsians, as first suspected by Osborn and his colleagues, since they bore unmistakable affinities to similar creatures of comparable age found elsewhere in China and North America.

But an even more extraordinary discovery emerged from Shabarakh Usu, one that would touch off frenzied media attention and, ironically, trigger nearly insurmountable hostility toward Andrews' expeditions among certain quarters in China and Mongolia. On July 13, while everyone was gathered for lunch, George Olsen casually announced that earlier that morning he had found fossilized eggs. "Inasmuch as the deposit was obviously Cretaceous and too early for large birds," Andrews said, "we did not take his story very seriously. We felt quite certain that his so-called eggs would prove to be sandstone concretions or some other geological phenomena. Nevertheless, we were all curious enough to go with him to inspect his find."

After carefully examining the objects, Granger was dumbfounded, announcing that the eggs were definitely reptilian and therefore must have been laid by dinosaurs. "The prospect was thrilling," wrote Andrews, "but we would not let ourselves think of it too seriously, and continued to criticize the supposition from every possible standpoint. But finally we had to admit that 'eggs are eggs,' and that we could make them out to be nothing else. It was evident that dinosaurs did lay eggs and that we had discovered the first specimens known to science."*

Olsen's historic discovery consisted of three partially broken eggs with fragments of others in the surrounding matrix. Several days later, Granger found a cluster of five more eggs, and Albert Johnson stumbled on a group of nine. Each was about eight inches long and seven inches in circumference. They were a reddish-brown color, elongated, and slightly flat. Their outer surface was striated, the interior quite smooth, and the shell was approximately one millimeter thick. Inside two of the broken eggs, the fossilized remains of tiny embryos were clearly visible. Several skulls of infants were uncovered, suggesting that they, along with other dinosaurs and unhatched nests of eggs, may have perished as a result of sudden drops in temperature or, more likely, of fierce sandstorms that swept the Flaming Cliffs, burying and choking large numbers of animals.

When Olsen cleared away the loose sediment from the ledge where the first eggs were embedded, he exposed the fragmentary skeleton of a birdlike dinosaur. It appeared to have died in the act of raiding a *Protoceratops'* nest. Osborn named the unidentified creature, found directly over the nest with a four-inch layer of sand between its skeleton and the eggs, *Oviraptor philoceratops*, literally, "egg thief that loves ceratopsians." After the block of matrix containing these discoveries was sent to the Museum, it was subsequently determined to contain thirteen eggs in two layers with the ends pointing toward the center, and there were indications that the nest originally held twenty eggs deposited in three layers. "This immediately put the [*Oviraptor*] under suspicion," wrote Osborn, "of having been overtaken by a sandstorm in the very act of robbing the . . . nest."

---

*Actually, the first suspected dinosaur eggs were found in the Pyrenees Mountains of France in 1859. A description of these fossils was published in 1877, but they went largely unnoticed at the time. In 1928, dinosaur eggs were unearthed in Montana, and since then they have come to light in India, Africa, China, the southwestern United States, Argentina, and elsewhere.

Oviraptorid, with a nest of eggs

Recent research has shown that several species of this creature inhabited the Gobi during late Cretaceous times, and that they belonged to a group of dinosaurs, known as theropods, that developed birdlike skeletal features and feathers prior to their extinction.

Altogether, around twenty-five or thirty eggs were collected in 1923, and there seemed to be no doubt they had been laid by *Protoceratops*. Their many skeletons outnumbered everything else at the Flaming Cliffs, but the dramatically beautiful formations yielded other species of Cretaceous denizens. Along with the grotesque *Oviraptor* lying above the first clutch of eggs, Osborn, writing in 1924, identified two carnivorous theropods—the deadly, swiftly moving *Velociraptor* (since made famous in the film *Jurassic Park*), with its razor-sharp, serrated teeth, powerful hind limbs, and knifelike talons; and a smaller predator known as *Saurornithoides*. There were also bones of a large dinosaur closely related to the fierce *Tyrannosaurus*; and Granger removed the skull of a species of crocodile later named *Shamosuchus*, which had probably inhabited a stream that once ran through Shabarakh Usu.

While paleontological and geological research continued at the Flaming Cliffs, Andrews and Mac Young journeyed sixty-five miles farther west to hunt bighorn sheep and ibex amid the peaks and ridges of Artsa Bogdo. Because the caravan was overdue and concern for its safety was mounting, Andrews left instructions that a messenger was to be sent if it appeared. When the caravan had still not shown up by July 30, Andrews and Mac returned to the main camp after a successful hunt. "I immediately rationed out remaining food, and reserved all the gasoline for emergency use in a single car. Since the flour was almost gone, by unanimous vote it was kept only for pasting fossils. We were reduced to tea and meat. Fortunately there were many antelopes, and we purchased several sheep from the Mongols. We got along well enough, our only real hardship being the lack of sugar."

Andrews managed temporarily to relieve the sugar shortage by purchasing two handfuls of coal-black chunks of a rocklike substance — purported to be sugar — from a caravan of Chinese traders. With the dubious-looking "pebbles," which at least tasted sweet, piled in the middle of a table, a ludicrous scene occurred when some of the world's most exalted scientists crowded around the table attempting to divide the little pile of black fragments into eight equal portions. "Granger ate his all at once," said Andrews, "but the rest of us spread our portions out over several days. Johnson decided he would make his into a sirup, but when the substance had been boiled and he saw the variety of insects, twigs, and others debris that floated to the surface, he admitted that 'where ignorance is bliss, tis folly to be wise.' I preferred to take the insects in a solid state and made my sugar into a round ball about the size of a walnut, which I could nibble at sparingly whenever I had a cup of tea."

Added to the shortage of flour, the burlap used for encasing fossils had long since been used up, leaving the men to resort to tent flaps, towels, washcloths, and finally miscellaneous items of clothing. "There is in the collection," Andrews wrote, "a beautiful dinosaur skull fortified with strips from my pajamas, and Frederick Morris, after considerable thought, presented one of his two pairs of trousers. That night Kaisen came in very much depressed, and, when I asked him why he looked so solemn, he said: '. . . I can use almost anything, but I simply cannot paste with Morris' pants.'"

As the days passed, Andrews became increasingly alarmed, and he began to contemplate the prospect of being stranded as winter approached. Finally, he decided to send two of his most trusted Mongols,

Tserin and Bato, on horseback to search for Merin. Within a few days, Bato returned with nothing to report. Tserin, meanwhile, attempted to retrace Merin's route eastward until feed for his horse became too scarce. Leaving it at a nomad encampment, he purchased a camel and rode for a week without seeing another human. Then he had the misfortune of encountering two lamas, who promptly attacked and robbed him. Tserin was so savagely beaten that he could scarcely make his way to a temple to recuperate from his injuries. When he was well enough to travel, he set out on a desperate journey back to the Flaming Cliffs, arriving nearly dead from starvation and thirst.

Meanwhile, Andrews, acutely aware of the seriousness of the expedition's plight, was about to risk using the last supplies of gasoline to lead a search party for the caravan. But before he could act, a wizened old lama unexpectedly appeared in camp. "Our Mongols greeted him with the greatest reverence," Andrews recounted, "and told us that he was a famous astrologer who had heard of our predicament and had come more than thirty miles to help us. The Mongols said that he would be able to tell us exactly where the caravan was. [He] made elaborate preparations and, after a long incantation, announced that the caravan was many days' travel away from us, but that we would hear definite news of it in three days. He said that our camels were dying and that Merin was having a very difficult time. Our Mongols believed him implicitly."

The old man's predictions were not far wrong. Word of Merin's whereabouts came four days later when one of Andrews' men discovered him sixty-five miles to the west near Artsa Bogdo. Finding it impossible to cross the sun-parched desert, he had circled far to the north where there was better feed. He had left some of the camels at wells along the way; others had died on the trail. Out of seventy-five camels, sixteen made it through carrying gasoline, food, and, to everyone's delight, sugar! Eventually, twenty-three more camels appeared with their loads intact.

With the caravan safely in camp, Olsen and Buckshot began packing the cache of fossils and zoological specimens that had accumulated over the past five weeks—no small task since they filled sixty empty supply boxes and gasoline tins and weighed five tons. Everything was meticulously cushioned in the thick hair shed by the camels, though they objected piteously when their loose winter coats were pulled off. By August 12, the expedition reluctantly left the Flaming Cliffs. Even as the procession was preparing to leave, fossils were still coming to light, and the cars were delayed for several hours at the last minute while Kaisen removed a superb *Protoceratops* skeleton lying on its stomach with outstretched legs.

On August 21, the men drove to the foot of Artsa Bogdo, its peaks already dusted with fresh snow, for three days of hunting before starting on the return trip eastward. Encountering a series of treacherous obstacles, the cars bounced over stretches of badlands, rocks, and deep sand, which wrecked a clutch on one of the Dodges and disabled a Fulton truck because of a broken pinion gear. Grossly overloaded, the vehicles could barely creep along through these obstructions under a blistering sun that continued to play havoc with Andrews' eyesight. There was genuine rejoicing when the scientists reached their former camp at Ula Usu.

Here, preparations were begun for what was to be a red-letter occasion for the expedition: a visit by Henry Fairfield Osborn himself. After months of planning, Osborn and his wife, Lucretia, were due to reach Shanghai the first week in September, then proceed by rail to Peking. Osborn's introduction to Mongolia would be the camp at Irdin Manha, the Valley of the Jewels, followed by a visit to Iren Dabasu.

Leaving Granger in charge of all arrangements, Andrews and Mac Young started for Kalgan on September 1. Arriving in Peking the next day, Andrews and Mac were met at the train station by a grim-faced Yvette and two army officers from the American legation, who bore grave news of the catastrophic earthquake, fires, and tidal waves that had devastated Yokohama and Tokyo the previous day. The fate of Osborn's ship, the *President Jackson*, supposedly anchored in Yokohama's harbor when the disaster struck, was unknown. Only after three frantic days of cables by the American minister in Peking was it learned that the Osborns' ship had sailed from Yokohama the day before the earthquake. In fact, they knew nothing of the disaster until they were informed in Shanghai by reporters and telegrams from the secretary of state and Andrews.

Osborn was enchanted by the expedition's headquarters in Peking. He wandered spellbound through its courtyards, laboratory, and art-filled rooms, asking a stream of questions about its history and former owners. While Lucretia stayed behind with Yvette, Andrews and Young whisked Osborn off to the desert the next day. Maintaining his customary sartorial splendor, Osborn was decked out in a hand-tailored bush jacket, knickerbockers, puttees, and a pith helmet.

"The trip was perfect," wrote Andrews. ". . . At four o'clock in the golden sunshine of a Mongolian afternoon, we saw [our] blue tents swimming in the desert mirage. They hovered and danced on the heat waves in the air, finally settling to earth like great blue birds as we neared the camp.

"It was one of the greatest days of my life and of the expedition when the man, whose brilliant prediction had sent us into the field, stepped from the car at our camp in the desert."

After tea, Osborn was driven to a pit where Granger had left a titanothere jaw partly exposed for his inspection. "This," Osborn exclaimed, "is the high point of my scientific life." But there was more to come. The next day was spent exploring fossil deposits around Iren Dabasu, "which have gone into history," Andrews remarked, "as the first identified Cretaceous strata [containing] dinosaur remains on the central Asian plateau." Osborn was particularly excited that five months before, Andrews had found the premolar of a *Coryphodon*, a member of a family of ungulates called Amblypoda. No other representatives of these creatures had been unearthed except in Europe. Returning to

Coryphodon

Prior to the discovery in Inner Mongolia of two teeth attributed to this animal, the only known specimens of Coryphodon—*an ungulate belonging to the Amblypoda family—were recovered from Eocene deposits in France and England.*

camp one afternoon, Osborn wandered off to prospect a small knoll about half a mile from the trail. Andrews waited in the car while Granger and the professor walked to the spot. "Osborn turned to me with a smile," Andrews remembered, "and said, 'I am going to find another *Coryphodon* tooth.' Two minutes later he shouted, waving his arms. 'I have it. Another tooth.'" Jumping out of the car, Andrews ran to the knoll, unable to believe his eyes as Osborn, beaming with delight, displayed a third upper premolar from the left side of a *Coryphodon's* jaw.

The last night in camp, Andrews and Osborn sat for an hour after dinner discussing the future of the expedition. Both men agreed that the challenge of exploring an immense area so laden with fossils and other scientific material could not be completed in the five years originally planned. "Ten years at least would be necessary," Andrews stated. "[It] meant that I must return to America to raise another quarter of a million dollars. Moreover, it was highly desirable for some of our staff to . . . study and evaluate the work already done. We determined, therefore, to declare a recess in the field operations for a year and start anew in 1925."

Everyone gradually drifted off to bed, only to be disturbed a short time later by the sound of galloping horses. Four men rode up armed with rifles. Obviously, they were bandits expecting easy pickings from defenseless travelers. Already alerted by a sentry he had posted, Andrews passed the word and the horsemen were quietly surrounded by the heavily armed explorers, who threatened to cut them down at the slightest move. After confiscating their weapons and placing them under guard, they were released unarmed the following morning. Later that day, it was learned from a colonel at a military garrison at Changpeh near Kalgan that the men were notorious brigands who had repeatedly escaped capture. A dozen soldiers were instantly dispatched on fast ponies; the bandits were caught before nightfall and shot. Osborn was elated: this last-minute taste of adventure, which could not have been more perfectly staged, was an ideal send-off for his final few hours in the Gobi.

Back in Peking, Osborn and his wife toured the Great Wall, the Ming Tombs, and visited most of the city's celebrated landmarks. They were feted with a constant round of luncheons, dinners, and receptions, and the professor was honored by various scientific societies. In mid-October, the Osborns, accompanied by Andrews, sailed for the United States from Shanghai. Yvette, pregant with her second child, remained at the Peking headquarters while Andrews busied himself with plans for another fund-raising campaign and a coast-to-coast lecture tour, although it meant a separation from his wife of almost nine months. Granger, meanwhile, sailed a few weeks later, bringing his Chinese assistants, Kan Chuen-pao, or Buckshot, and Liu Hsi-ku, who had received special visas to spend the winter studying paleontological conservation techniques at the Museum in New York. Docking briefly with the Osborns at Yokohama and Tokyo, Andrews was shocked at the devastation wrought by the earthquake six weeks earlier:

> I could hardly believe that such complete destruction of any city was possible. Not a single undamaged building stood in Yokohama. The city looked like the dump heap in a brickyard. Salvage had only begun. I stood on the ruins of the Grand Hotel and saw men take out the charred body of one of my intimate friends from a crushed bathtub. . . . I gazed sadly at the heap of ruin which had been Number Nine, knowing that somewhere in that pile of debris were the ashes of Mother Jesus, for, although many of the girls were saved, no trace of their mistress was ever found.

CHAPTER 13

When news of the dinosaur eggs finally hit newspapers around the world, the reaction to the announcement was phenomenal. The press and public alike devoured the discovery with a frenzy that even Andrews found difficult to fathom. Minutes after his ship docked at Victoria, reporters from virtually every major West Coast city swarmed aboard. Rushing to beat out competitors, a representative of a Seattle paper offered $1,500 for exclusive use of the dinosaur egg photographs for one week. Another reporter upped the ante to $3,000, and a journalist from the *San Francisco Chronicle* raised it to $5,000. In Seattle, the mayor, the president of the Chicago, Milwaukee, and St. Paul railroad, which would transport Andrews and the Osborns across country, and a host of other dignitaries were on hand to meet the ship. Reporters, excited crowds, and popping flashbulbs greeted Andrews at every stop across the country.

His first lecture at the American Museum was jammed. Four thousand people—including John D. Rockefeller, Jr., with two of his children, and the renowned Egyptologist James Breasted—tried to cram into an auditorium designed to hold an audience of fourteen hundred. To solve the problem, Andrews spoke for an hour, then repeated the lecture a second time. A few days later, Rockefeller came to the Museum to see the eggs, "and found the hall . . ." Andrews noted, "packed almost like a subway train."

During eight months in America, he complained, all he heard was "Dinosaur eggs! Dinosaur eggs! Vainly did I try to tell of the other vastly more important discoveries of the expedition. No one was interested. No one even listened. Eventually, I became philosophical about it."

Yet dinosaur eggs were not the only attraction. The man behind their discovery was equally riveting to masses of admirers. Andrews had come to epitomize a swashbuckling adventurer whose determination had forced the Gobi to give up its ages-old secrets. An incongruous mixture of explorer, socialite, scientist, sportsman, author, lecturer, and diplomat, Andrews' fame had now soared to extraordinary heights. Along with a deluge of newspaper articles recounting his explorations, he was featured in periodicals such as the *Saturday Evening Post*, *World's Work*, the *Illustrated London News*, *Harper's*, the *Literary Digest*, and *Cosmopolitan*. In the August, 1924, issue of *Asia*, its editor, Louis Froelich, penned a tribute to his friend entitled "Andrews of Mongolia," which summarized his achievements in wildly laudatory terms. But his celebrity status was firmly assured when a portrait of Andrews appeared on the October 29, 1923, cover of *Time* magazine.

Influenced by the outpouring of popular interest in the expeditions' discoveries, Andrews decided to raise his financial objective to $300,000. With his entrepreneurial instincts, he recognized that the dinosaur eggs offered a golden opportunity to solicit money from an untapped source: the general public. Impressed with the explorers' achievements, some of Andrews' original backers—Morgan, Rockefeller, Baker, and others—agreed to increase the amount of their pledges. But Andrews was now casting about for a means to induce the average person to contribute, and he felt the eggs provided an ideal "gimmick." Already, the eggs had acquired a folklore of their own, and newspapers and magazines were inundated with letters, satirical pieces, and poems on the subject.

While in New York, Andrews was living at Osborn's house at 998 Fifth Avenue, and during breakfast one morning, he unveiled a strategy he had been mulling over for some time. He was convinced, he said, that the public at large would help finance the expeditions, ". . . but they think small contributions aren't wanted. They believe this is only a rich man's show. If one of the dinosaur eggs was auctioned off with the money going to the expeditions' funds, it would be a grand publicity stunt. Every news story could explain that we've got to have money or quit work; [and] small contributions are welcome."

Thus began the "Great Dinosaur Egg Auction." On the morning of January 7, 1924, about forty reporters crowded into Andrews' office at the Museum to hear details of the scheme. The egg would go to the highest bidder, but Andrews asked that all articles concerning the auction stress that donations of any size were needed. Had they foreseen the consequences of this innocuous-seeming ploy, Andrews and Osborn would have rejected the plan out of hand. "Nothing else so disastrous ever happened to the expedition," Andrews later admitted.

By the following day, major papers throughout the country ran articles and editorials about the auction, thereby touching off a rash of publicity that included features in the *New York Times,* the *New York Herald,* the *Washington Post,* the *Chicago Tribune,* the *Boston Globe,* and the *New York Evening Post.* Even though the bidding was open to anyone, letters announcing the auction were sent to specially selected prospects: the British Museum, the Carnegie Museum, the Smithsonian Institution, the *Illustrated London News,* and the National Geographic Society, to name but a few. Bids were received from a number of institutions, though the egg was eventually purchased by Austin Colgate (Bayard Colgate's uncle) for $5,000 as a gift to Colgate University.

But the stunt backfired. "Up to this time the Chinese and Mongols had taken us at face value," Andrews wrote. "Now they thought we were making money out of our explorations. We had found about thirty eggs. If one was worth five thousand dollars, the whole lot must be valued at one hundred fifty thousand. They read about the other fossils — dinosaurs, titanotheres, *Baluchitherium.* Probably those, too, were worth their weight in gold. Why should the Mongols and Chinese let us have such priceless treasures for nothing?" It was a dilemma that would create continual problems for the expeditions, far outweighing the benefit of proceeds from the auction.

The affair did have its bright side. Just as Andrews had hoped, small donations began arriving from all over the country in $5, $10, $25, $50, and $100 amounts (one check was for $10,000). Ultimately, the publicity surrounding the auction garnered more than $50,000 in contributions from thirty-six states and eight foreign countries. Nor did Andrews hesitate to resort to blatant chauvinism to stir up public support. In a fund-raising letter to members of the American Museum, he flaunted:

This is an all-American Expedition. We are carrying American ideals, American science, and the American flag into one of the least known countries of the world. We are bringing back the fruits for the entire American people.

Thus far the expense has been borne almost entirely by generous men and women of New York. But now I must go outside that city to ask for support. . . . We need the support of every man and woman who is interested in the advancement of American exploration and American science. Will you help us? . . . It is *your* expedition and I feel sure that you do not wish the work to end when it has been so successfully begun.

Along with the rush of small donations, there were more lucrative temptations. A novelty manufacturer, for example, proposed "flooding the world" with casts of dinosaur eggs as paperweights, each one signed by Andrews, but he rejected the idea as "science camouflaging business." Both the *New York Times* and the *New York Herald* came forward with bids for exclusive rights to the expeditions' news releases, which would then be syndicated to other papers, with proceeds from syndication going to underwrite the project. It was an enticing offer, but Andrews feared serious repercussions. Since the American Museum was a public institution, partially tax-supported, its press releases could not be made available on a restricted basis. Inevitably, a representative of the media giant William Randolph Hearst appeared with a proposition that would have instantly eliminated any further need for fund-raising. Hearst was prepared to offer $250,000 outright for exclusive newspaper, magazine, and motion picture rights to the expeditions. "There was just one answer for that," Andrews proclaimed. "No! My mind was made up before his representative had ceased speaking."

Andrews felt no such hesitancy when it came to trading advertising for equipment and supplies. He willingly entered into an agreement with Dodge Brothers to provide a new fleet of automobiles for the 1925 expedition. Always in the past, Dodge had refused to contribute its cars for such ventures. Impressed by their performance in the Gobi and the publicity Andrews had generated by driving Dodges across such forbidding country, the company decided to make an exception to its usual policy. It offered to supply five automobiles—four open-bodied vehicles and one regular touring car, used by Andrews for reconnaissance.

All were specially equipped with extra-heavy springs, iron bumpers with hooks designed to be hauled by a winch, oversized gas tanks, and two spare tires. In turn, Andrews filled his press releases and magazine articles with enthusiastic praise for the automobile's sturdiness.

Soon he had worked out similar arrangements with numerous companies: Eveready flashlights, Smith-Corona typewriters, MJB coffee, Savage Arms, Royal Cord tires (made by U.S. Rubber Company), and an array of other products were supplied to the expeditions in return for advertising. In resorting to this approach, Andrews pioneered the concept, so commonplace today, of corporate sponsorship for scientific ventures. "[It] was entirely legitimate," he insisted. "We simply endorsed publicly, products which had proved their worth. . . . There were, however, critics who said we had sold out to Dodge and [Standard Oil]. But it didn't bother me in the slightest."

On December 12, 1923, Andrews inaugurated a lecture tour before a glittering crowd at Town Hall. The evening was under the patronage of a select group of Andrews' original sponsors, including among others Childs Frick, W. Averell Harriman, J. P. Morgan, Jr., Sidney M. Colgate, John D. Rockefeller, Jr., Thomas W. Lamont, Arthur Curtiss James, John T. Pratt, Dwight W. Morrow, Charles L. Bernheimer, Albert Wiggin, and their wives, along with George F. Baker and Mrs. Willard Straight. Using hand-colored slides taken in Mongolia, Andrews deftly guided his audience through an evening of exciting adventures and discoveries. It was a runaway success!

Earlier, the American Asiatic Association had negotiated a contract with the lecture management firm of Louis J. Alber in Cleveland, Ohio. It called for a grueling speaking tour that would take Andrews to dozens of cities across the country. Net profits from his engagements were to be divided, with one-third each going to Andrews, the Alber agency, and *Asia* magazine. So great was the demand to hear Andrews that he lectured virtually every day, often two or three times when he was not traveling. During four months on the road, he appeared before nearly 125 audiences, and when the tour ended, he complained of feeling like a "sucked orange."

In Los Angeles, he was taken in hand by the legendary John Barrymore, an avid fan of Andrews' exploits. Barrymore, then at the height of his fame for his performance in *Hamlet,* delighted in introducing the renowned explorer to a host of Hollywood's celebrities and showing him the city's attractions. Barrymore was obsessed with the idea of owning a dinosaur egg, and although that was not possible, Andrews presented him with a dozen bits of shell about the size of a thumbnail, along with a letter of authentication.

While earning a substantial sum for the expeditions, Andrews' lectures had the unfortunate effect of rekindling the debate over evolution. Some of his appearances, especially in the South and Midwest, were picketed by Fundamentalists; others bombarded him with derogatory letters. Under a headline in the *Scarsdale Inquirer* stating, ROY ANDREWS A PERIL TO YOUTH, a local resident objected "to having anybody talk . . . who will spread among the children and younger folk . . . the dreadful doctrine of evolution." He was particularly incensed because "among the pictures which Mr. Andrews is to show are some fossilized eggs laid by reptiles '[millions] of years ago.'" According to the writer, the man who does that "insults our intelligence and our God by claiming to act as an authority for the age of the earth, which is written in the upper left hand corner of my Bible as 4004 B.C."

Regardless of the fact that the Missing Link had eluded the fossil hunters, Andrews and Osborn held to their conviction that Asia was the "Garden of Eden," anthropologically speaking, and that man's ancestors would ultimately emerge from its unexplored depths. Not unexpectedly, with the Scopes Monkey Trial set to begin in July 1925, William Jennings Bryan again went on the attack. Venting his notorious wrath, he characterized Osborn as "a tall professor who comes down out of the trees to push good people who believe in God off the sidewalks." The issue soon flared into a repetition of the famous Osborn-Bryan debate of 1922 in which the two adversaries had flailed each other in the press and from lecture platforms. By 1925, the *New York Times* was providing a widely read forum for the renewed controversy, with undisguised bias favoring Osborn's proevolution stance. It willingly allowed excerpts from his stinging critiques of Fundamentalist ideology—going back to his 1922 clash with William Jennings Bryan in the *Times*—to be published in his book, *The Earth Speaks to Bryan*, which appeared at the start of the Scopes trial.

While this contentious war of words over evolution raged on, Andrews went forward with plans for his 1925 expedition, the most ambitious and costly of them all. He was in boundless high spirits—and with good reason. It had been one of his best years, both professionally and personally. He had raised approximately $265,000, plus another $20,000 that the Field Museum in Chicago had paid for duplicate specimens from the Gobi, including dinosaur skeletons and several eggs.

On July 18, 1923, Osborn decided to appoint a committee to supervise the publication of the expeditions' scientific results, which were beginning to appear regularly in the Museum's *Novitates* and *Bulletin*, apart from an ever-expanding list of popular publications. By now, Andrews' salary had also been raised to $4,600 a year, and in 1923 alone, he received $2,800 for his share of articles for *Asia* and proceeds from lectures.

On January 20, 1924, Yvette gave birth in Peking to a boy— Osborn's godson—who, after months of discussion between Andrews, Yvette, and Osborn, was named Roy Kevin Andrews. "A most adorable young man," declared Yvette, ". . . with the bluest of blue eyes," added Andrews, "and the merriest of dispositions." It had been "a terrible wrench to leave Yvette [during her pregnancy]," Andrews had confided in a letter to his aunt, "but it simply could not be helped."

Arriving in Peking on July 4, 1924, Andrews immersed himself in organizing the upcoming expedition. Because its scientific scope was to include studies of the Gobi's extinct flora, ancient climate, and archaeology, plus a more detailed mapping program, five extra specialists were added to the staff. The photographer J. B. Shackelford rejoined his former companions after spending most of 1923 in Tahiti as a cinematographer for William Randolph Hearst's motion picture company. "I hope," Andrews had written to Granger, "that Marion Davies [Hearst's film-star mistress] will not get such a stranglehold on Shack that he won't come with us." He needn't have worried. Shackelford appeared in Peking exuding more enthusiasm at the prospect of returning to the Gobi than he ever expressed for the famously seductive Marion Davies.

Added to the veterans—Granger, Berkey, Morris, Olsen, Shackelford, and Mac Young (Kaisen and Johnson were unavailable that year)—the Carnegie Institution in Washington sent over Ralph W. Chaney, an expert paleobotanist with extensive training in geology, living plants, and zoology. As his chief topographer, Andrews engaged

Major L. B. Roberts, a member of the Army Reserve and a close friend of Charles Berkey. Rounding out Roberts' mapmaking team were Lieutenants Frederick B. Butler of the Army Corps of Engineers, and H. O. Robinson, an Englishman assigned to the First Royal Lancashire Regiment at the British legation in Peking. Another Britisher, a master mechanic named Norman Lovell, was hired to help Mac Young maintain the cars and trucks. For the first time, Andrews took along a surgeon, Dr. Harold Loucks, on leave from the Peking Union Medical College. And the herpetologist Clifford Pope, who had spent 1922–1923 collecting reptiles, amphibians, and fish on Hainan Island, returned to Peking in July of 1925 to prepare for a sojourn to the provinces of Fukien and Kiangsi.

The Swedish-born archaeologist Nels C. Nelson was also a newcomer to the group. Having joined the American Museum in 1912, he was an authority on European prehistory. Nelson, an archconservative in scientific matters and nearly everything else, sported a glass eye, an affliction that gave rise in the expedition's lore to two versions of an amusing incident with Mongol bandits. According to one story, Nelson and several companions suddenly found themselves encircled by brigands while they were engrossed in collecting artifacts. Unable to get to their weapons, the quick-thinking Nelson popped out his glass eye with a great flourish as a demonstration of the white men's magical powers, causing the Mongols to flee in abject terror. Another account has it that Nelson placed the eye on a rock, calmly proceeded with his excavations, and told the Mongols through an interpreter that his eye would be watching for any suspicious moves, a strategy that evoked the same desperate fear of the terrifying "evil eye."

Augmenting the expedition's technical staff were Buckshot and Liu, the two Chinese whom Granger had been training as collectors. At the instigation of Osborn and the Chinese Geological Survey, they had spent the winter of 1924 in New York studying at the Museum's paleontological laboratory. But the experience was not without serious drawbacks: Liu and Buckshot hated New York, could not eat the food, and became inconsolably homesick. Liu had to be fitted with eyeglasses, and both men suffered constantly from stomach disorders and colds. Overjoyed to return to Peking, they had cost the Museum a substantial sum of money, although Granger admitted their training made them invaluable assistants in the field.

The 1925 expedition was the grandest and most ambitious to date: fourteen scientists and technicians; eleven Chinese cooks, taxidermists, camp assistants, and chauffeurs; fourteen Mongol interpreters and car-avanners; one Dodge touring car and four three-quarter-ton open cars; two one-ton Fulton trucks; and a caravan of approximately 125 camels. It carried four thousand gallons of gasoline, one hundred gallons of oil, several dozen extra tires; two and a half tons of flour, a ton of rice, half a ton of sugar, and an equally prodigious amount of other staples and special equipment.

As Andrews had anticipated, the worst difficulties again arose in try-ing to deal with Mongol officials. By now, Urga was a cauldron of Bol-shevik ideology and political reform. Buriats and Soviet Secret Police maintained a climate of fear and oversaw every official agency. For-eigners were viewed with even more suspicion than Andrews had expe-rienced in 1922, and their passports and luggage were examined repeatedly. On May 20, 1924, the Hutukhtu had died, and the Com-munists decreed that no successor would be recognized. Everything was under the direct control of Moscow, and foreign businesses were decidedly unwelcome. A "Travel Pass" was required to leave the city; it cost $120 and often took two or three days to obtain. All members of the pre-Bolshevik regime were rapidly being replaced, executed, or sent into exile, and their places taken by Russians or Buriats.

On August 25, 1924, Andrews arrived in Urga for almost a month's stay to finalize arrangements for the expedition. He was accompanied by F. A. Larsen, whose influence was now sadly weakened, one of his mechanics, Norman Lovell, and Gordon Vereker, first secretary of the British legation in Peking, who was traveling on unrelated business. As expected, the negotiations quickly disintegrated into the familiar litany of accusations that the expedition was engaged in spying, prospecting for minerals, and stealing priceless fossils. Andrews now found it diffi-cult to refute these charges in view of the dinosaur egg auction, which caused several Mongol officials to explode with rage and claim that the project was nothing more than a commercial venture.

Once again, however, it was T. Badmajapoff, now a high-ranking government minister, who saved the situation through deft negotiations with the premier, a reasonable man named Tserin Dorchy, and the ministers of foreign affairs and education. Although Mongolian territory was now closed to all foreign scientists, Andrews argued that he was still operating under his agreement reached in 1922.

> It was a delicate business [Andrews recalled]. . . . I found that a Scientific Committee had been formed, apparently for our benefit. Its object appeared to be to obstruct our scientific work as much as possible. . . . Then the Secret Service took a hand and said that we could not be allowed to make maps or do any topographic work whatsoever . . . but, guided by the astute diplomat, Badmajapoff, I made considerable progress. At the end of three weeks the authorities agreed to permit the Expedition to work in Outer Mongolia under certain conditions. These included the presentation to Urga of a duplicate set of all our collections, publications and maps [which, in fact, were being prepared for shipment from New York and Peking and included material from the 1922 and 1923 seasons]. . . . Moreover, we must take with us two Mongolian representatives, one from the Scientific Committee and the other from the Secret Police. These gentlemen were to act as the recording angels of our actions. I was to pay their expenses and salaries as well. The passports for ourselves, cars, camels and equipment cost three thousand dollars. It was agreed that they would be sent to me in Peking.

Andrews returned from Mongolia on September 20, dubious that any of these agreements actually would be carried out. But to his relief, a letter from Badmajapoff arrived on December 13 announcing that all arrangements were complete and the passports were on their way to Peking. As before, Badmajapoff's friendship with Andrews and belief in the validity of his scientific objectives had made the Central Asiatic Expeditions possible, despite roadblocks no one could have foreseen.

CHAPTER 14

Armed with a sheaf of permits from authorities in Urga, Merin, his caravanners, and 125 camels left Kalgan on February 20. In a jubilant mood, the drivers pierced the morning silence with a shrill chant as they rode away into a thick mist that had settled over the desert. Andrews' instructions were to proceed by slow marches and rest the camels wherever feed was plentiful. The entire expedition was to rendezvous at the Flaming Cliffs by May 1. For Merin, this meant a journey of eight hundred miles across long stretches of difficult terrain.

On April 18, Roberts, Butler, and Robinson set up their transit, plane table, and stadia rods at the railroad station at Kalgan. Working from this known elevation point, they gradually extended their topographical measurements through the Wanchuan Pass to the village of Miao Tan and onto the plateau, where they were to await the main party's arrival. Ironically, the district around Kalgan was now controlled by Andrews' adversary Marshall Feng Yu-hsiang, the "Christian" general with whom he had clashed in 1922 during his hunt for *takin* in Shensi Province. Feng's troops were hastily smoothing deep ruts in the otherwise impassable road out of Kalgan so that truckloads of soldiers could presumably chase down bandits who were causing havoc in the region. One car after another was being attacked, and three American merchants had recently lost $80,000 worth of sable skins to brigands. The situation was so precarious that the foreign commissioner at Kalgan—who first tried to discourage the expedition from continuing—had demanded that Andrews sign a legally worthless document releasing the Chinese from all responsibility for the explorers' safety.

But the pursuit of bandits was not Feng's only motive for repairing the road. Only later was it learned that he was using the Kalgan-Urga route to secretly smuggle arms and ammunition from Russia, which supported Feng's ambitious plot to destroy Chang Tso-lin's army and gain control over northern China. Feng, in turn, had promised the Soviets access to the warm-water port at Tientsin if his scheme succeeded. (Russia's only Pacific port of any size at Vladivostok was ice-bound during the winter.) The stakes, therefore, were extremely high, and Feng was alarmed that Andrews' massive expedition might discover his covert operation with the Russians. Yet there was no way Feng could stop an expedition bathed in the glare of international publicity without arousing suspicion, so it was begrudgingly allowed to proceed.

On the evening of April 27, while camped near a shallow stream known as Wolf Creek, a Mongol appeared on camelback. He brought alarming news from Merin. Fifty miles from Ula Usu, guards at a military post, or *yamen*, had confiscated the caravan. Having found ammunition in two of the boxes, along with other "contraband," the camels had been held for a month without adequate food, and the Mongol reported that many of the boxes had been torn open. Aware that Merin carried all necessary permits, Andrews suspected that the Buriats in charge of the *yamen* were attempting to extort "squeeze" before releasing the caravan. "The detention of our caravan," he fumed, "was a serious blow to all our plans. Instead of having a base at Shabarakh Usu, with rested camels awaiting our arrival, they had covered only half the distance, with four hundred miles of the worst desert [still] to cross. It meant that I should have to readjust the entire work of the Expedition."

After moving the camp to Ula Usu three days later, a lama arrived with news that the *yamen* where the caravan was being held was heavily guarded, and orders had been issued that Andrews was to be taken to Urga and shot. Dismissing the threat as "a most interesting prospect," Andrews started for the *yamen* early the next morning with two cars and eight men—Granger, Loucks, Shackelford, Lovell, Young, Buckshot, and two Mongols. "All were armed and determined to have the caravan," Andrews declared, "with or without a fight." Fifty-two miles up the trail five Mongol soldiers, sporting bandoliers, Mausers, and rifles, awaited the explorers. "We unceremoniously bundled one of them into a car as a guide to the *yamen*, twenty-five miles ahead." It consisted of two yurts surrounded by six tents. Merin and his drivers were overjoyed at the sight of the automobiles and the fact that their occupants, carrying a veritable arsenal, had obviously come to rescue them. Andrews and his companions were greatly relieved to find that the damage to the boxes was not as severe as anticipated, although a few had been opened and thoroughly searched.

Moments later, an arrogant Buriat officer appeared. He ordered Andrews to surrender his rifle and revolver and prepare to leave for Urga at once. Andrews was informed that the commandant would send for him in due time, while the other men remained under guard pending further instructions. Enraged almost beyond control, Andrews retorted, "Tell your chief that *we* are ready to see him *now*." Leaving Shackelford and Loucks to guard the cars, Andrews strode angrily toward the largest yurt, threw open the door flap, and stepped boldly inside followed by Granger, Lovell, and Young, Buckshot, and his two Mongol interpreters, Tserin and Aiochi. About twenty soldiers and lamas were seated in a circle. "Who is the head man?" Andrews shouted. A lama at the back of the yurt, attired in a gorgeous yellow silk robe and a sable-trimmed hat, slowly raised his hand. Andrews demanded to know why the caravan's permits had been ignored. Visibly intimidated by Andrews' undisguised hostility, the lama replied in a barely audible voice that shotgun shells were found in two of the boxes. He stammered that he had wanted to allow the caravan to pass anyway, but the military officers at the *yamen* refused because the crates also contained "bombs"—in reality, Eveready flashlight batteries—two Chinese bayonets used for digging out fossils, and a collection of seditious literature: *Asia, World's Work, Saturday Evening Post, Outlook,* and other "dangerous" magazines.

Andrews reminded the lama that his permits covered everything transported by the caravan; the soldiers, he loudly proclaimed, had behaved like brigands. They had no authority to inspect any of the boxes, and he threatened to forcibly take the lama and his military cohorts to Urga to answer to government authorities. In an outburst of frustration, Andrews suddenly brought his fist down full force with an ear-shattering crash on a sheet metal stove, causing the lama to break a string of yellow prayer beads he was nervously fingering. By now, confused and psychologically outmaneuvered, the Mongols hastened to release the caravan.

The next day, Andrews removed extra gas and provisions to be stored at a nearby lamasery for the return trip to Kalgan, and the caravan left the *yamen* in the dusty morning haze. But the damage had been done. It would never be able to reach the Flaming Cliffs by May 1 as originally planned. "I told Merin to travel fast," wrote Andrews, "leaving the weakest camels by the trail. We must sacrifice the animals in order to make up as much as possible of the lost time. Merin thought he could reach Shabarakh Usu in three weeks, and I gave him twenty-five days maximum."

From Ula Usu, the scientists' route carried them south by southwest past a series of wells and nomad encampments. Along the way, the paleontologists investigated promising formations, finding bone fragments and occasional complete skeletons. Ralph Chaney, the botanist, gathered fossilized plants in shale, tamarisk cuttings, and a variety of shrubs, grasses, leaves, and other flora that might provide clues to the Gobi's vegetation and climate over aeons of time. Lovell picked up a handful of iron arrowheads probably used by Mongol warriors. The topographers devised a method of running their surveys from one point to another with remarkable accuracy by using the automobiles' fenders, hoods, or windshields to align their sightings, then measuring the height of each focal point above the ground to calculate topographical intervals. Andrews, meanwhile, trapped and shot animals and birds, which the Chinese taxidermists skinned, preserved with chemical compounds, labeled, and packed in boxes. Now and then, the plains thundered with thousands of antelope, gazelle, and wild asses, though Andrews killed only a few exceptional examples to supplement his already substantial collection.

Once they reached the Flaming Cliffs, elated to be back and give the expedition's new members a chance to see its spectacular formations, everyone set out to reexplore the area. On the plains near the cliffs, Nels Nelson, the archaeologist, began finding a remarkable array of man-made artifacts in the form of hammerstones, axes, scrapers, knives, arrowpoints, and blades made of red jasper, chalcedony, chert, and other stones. Even a few fragments of *Protoceratops* and extinct ostrich egg shells had been skillfully drilled with small holes to make beads. Associated with the oldest of these objects were large stone cores from which the implements had originally been fashioned; and there were remnants of fire pits containing ashes and bits of charcoal. Some of the hammerstones and axes were similar to Azilian artifacts from late Paleolithic sites in Europe, dating from about twenty thousand years ago. But the existence of pottery fragments among the items at Shabarakh Usu confirmed that many of these implements were no older than the late Neolithic period. The same was true of the delicately worked arrowheads found in such abundance, which were unknown in Paleolithic times, although viewed in context, the full inventory of objects appeared to show a progression from late Paleolithic to Neolithic technology.

All of these artifacts—and there were thousands of them—were picked up on flat surfaces between sand dunes, where swirling winds had blown away the sediments exposing the implements. As a result, Nelson, Andrews, and Granger called the unknown people who manufactured these objects (which have still never been systematically studied) "Dune Dwellers" for lack of a more definitive terminology. Assuredly, they were far removed in time from Osborn's early anthropoids, even though he and Andrews tenaciously held to the theory that such creatures had inhabited Asia hundreds of thousands of years earlier.

Every day, the Flaming Cliffs continued to yield its inexhaustible fossils. Immediately after arriving, Buckshot and Liu each found wonderfully preserved *Protoceratops* skulls. George Olsen, the expedition's "champion dinosaur egg hunter," extricated five nearly perfect eggs under a fifty-pound rock (they now reside in the Field Museum of Natural History in Chicago); and just before leaving Shabarakh Usu for the last time, Olsen uncovered an even dozen eggs, several of which were the finest examples ever retrieved from the area. Norman Lovell discovered eighteen eggs standing on end in a double row; and Robinson stumbled upon a cache of egg fragments, larger and rounder than those attributed to *Protoceratops*, laid by a giant ostrich known as *Struthiolithus*, which had inhabited the Gobi much later during the Pliocene or early Pleistocene epoch. In one location, Chaney collected 750 pieces of fossilized eggshells—and so it went day after day as the Flaming Cliffs revealed their scientific marvels.

Of all the Gobi's wonders, this spot was far and away Andrews' favorite: its towering rocks so dramatically sculpted by centuries of wind and rain, its brilliant colors contrasting sharply with the sandy plains that encircled the basin for miles. "Like a fairy city it is ever-changing," wrote Andrews. "In the flat light of midday the strange forms shrink and lose their shape; but when the sun is low the Flaming Cliffs assume a deeper red, and a wild mysterious beauty lies with the purple shadows in every canyon." Andrews liked nothing better than to sit in front of the tents in the evening, surrounded by his companions, smoking and listening to recordings of Caruso on their Sonora phonograph, and contemplating the steadily changing spectacle of colors and shapes.

Despite the twenty-five days Andrews allotted Merin to reach Shabarakh Usu after leaving the *yamen* where he had been held captive, the caravan was two weeks late. As usual, Andrews feared the worst. But one night, he recalled, "we heard a wild Mongol song in the moonlight. It was answered from camp and everyone ran out in pajamas to see Merin silhouetted against the sky on the rim of the basin. Eighty-six camels were close behind him and the caravan was safe." He had found the journey even more daunting than expected and lost numerous camels because of drought and poor grazing. For that matter, the explorers had experienced their own share of rough going. Since leaving Ula Usu, they had endured fierce winds, extreme changes in temperature, and explosive sandstorms, two of which kept them in their tents day and night, their faces covered with damp cloths to combat the suffocating dust.

Now, however, they were suddenly confronted by another threat. A messenger arrived from Urga summoning Andrews to come at once. "New government regulations had been passed," he was told, "which required discussion if we were to continue our explorations." From past experience, he knew the situation did not bode well.

Andrews and Young left for Urga on May 24, taking one car and their reliable assistant and interpreter Tserin. As they drove out of camp, an icy gale was whipping up a yellowish-brown wall of sand and gravel ahead of a storm. By seven o'clock that night, they stopped at the yurt of hospitable nomads, who built a blazing *argul* fire to warm their visitors. Andrews and Young erected their tent just outside the yurt with the back side facing the blowing storm, but during the night, the wind shifted directions. In the morning, the tent was filled with snow. "All our clothes," Andrews wrote, "were deeply buried . . . [and the car's] engine was as solidly packed as though we had shoveled snow inside."

Getting under way at eleven o'clock the next morning, the rest of the three-hundred-mile drive was a laborious struggle against blowing drifts and hidden ravines filled with six to eight feet of snow. Time and again, the automobile had to be dug out, the wheels jacked up, and rocks piled end-to-end to make a crude roadway over which the car climbed out of the slippery pitfalls. Because Andrews had injured his right collarbone in a New Year's Day steeplechase and could barely move his arm, Young and Tserin were saddled with the heavy work while Andrews gathered rocks and drove. After spending a second snow-bound night in another yurt—which they shared with a foul-smelling goat, calves, a baby, and two adults—the following morning was bathed in brilliant sunlight that illuminated the colorful temple roofs and golden cupolas of Urga like "a beautiful jewel," Andrews observed, "set in the green valley of the Tola River. . . . Peaceful enough it looked in the spring sunshine, but I knew that it was a city of suspicion, and one not to be entered without due thought of how one was to get out."

The tribulations Andrews and Young had feared were inflicted on them from the day they arrived. Everything appeared to begin well enough when they were taken to a ramshackle museum to help install a *Protoceratops* skeleton, a nest of dinosaur eggs, a reproduction of *Baluchitherium*, and other scientific specimens provided by the American Museum in accordance with Andrews' original agreement signed in 1922. As this work progressed, they were subjected to a "merciless searching of our persons," Andrews complained bitterly, "and every scrap of writing and all books were sent to the Secret Service office to be perused by censors."

Next came a shattering blow. Andrews was coldly informed by Bolshevik officials that all arrangements for his explorations negotiated in 1924 had been nullified by "new regulations" passed a month earlier for "our especial benefit." Among a list of impossible demands, "we were required to bring all our collections to Urga, where the Scientific Committee was to take what it wished. No maps of any kind could be made and no geological work done. Moreover, the Mongol government was to designate two students who were to be sent to America and educated at Harvard University entirely at the Expeditions' expense." Henceforth, no radios or wireless equipment could be used, leaving the explorers cut off from time signals and news. And at least two government agents were to accompany the expedition, with complete license to send back reports on its activities. Andrews was positive such restrictions reflected a smoldering resentment over his auction of the dinosaur egg, and represented a clumsy attempt to prevent anything of worth, particularly fossils with artificially inflated values, from leaving the country.

Days of negotiations were required before the most absurd of these demands could be ironed out. "In the meantime," Andrews wrote, "we were under constant espionage. Tricks of various sorts were resorted to, whereby the authorities hoped that we might do, or reveal, something which would show that our scientific work was merely a camouflage for some sinister design." As had so often happened in the past, T. Badmajapoff ended up as a decisive player in the farce. Having eliminated or reached a compromise on the most objectionable features of the government's demands, they signed a new set of documents on June 4. Andrews had no objection to the revised permits other than the presence of the two government agents, who were to be taken along at the expedition's expense as representatives of the Scientific Committee.

While awaiting a resolution of the complications in Urga, Andrews and Young engaged in a risky episode that might easily have had disastrous consequences. Tradition required that the corpse of the Hutukhtu, or Living Buddha, who had died on May 20, 1924, must lie in state for a year before being permanently entombed in a temple. Because the Communists had outlawed the installation of a successor as part of their campaign to eradicate Lamaism, hordes of pilgrims were streaming into Urga for the burial rites of their last spiritual leader. Andrews and Young were eager to witness the ceremony, but the presence of foreigners at such a sacred event was strictly forbidden under penalty of banishment from the country or death.

A young Mongol named Dalai Badmajapoff, apparently unrelated to Andrews' friend T. Badmajapoff, offered to conspire with Andrews and Young to gain their entrance to the temple. Using a ruse that involved staining their hands and faces with brown color and dressing in Buddhist robes and hats, the disguise was so convincing that it allowed them to melt unnoticed into the throngs of other lamas. Why Andrews would have jeopardized the expedition's integrity by taking part in such a foolish prank is unclear. Had he and Young been caught disguised as monks at a holy ritual forbidden to outsiders, their antagonists would surely have had cause to expel the expedition, seize its collections and equipment, and hand out whatever punishment they felt Andrews and Young deserved. It was exactly the kind of adventure Andrews found irresistible, and Mac Young's devil-may-care attitude made him an equally enthusiastic accomplice.

> To a priest at the door [Andrews recounted] Dalai explained that Mac and I were Mongols from the Alashan Desert who had journeyed to Urga, like many other pilgrims for this ceremony; that we were both deaf and dumb, a not very uncommon thing in Mongolia. The priest welcomed us courteously and we passed inside. I was a little worried about my blue eyes—Mac's were gray—but the temple was lighted only with candles and in the half darkness they got by.
> . . . At the far end of the room, facing the entrance, sat my old friend the Hutukhtu on a golden throne. He looked natural enough to speak except that he was only half his normal size and was completely gilded with gold leaf. His features were perfect. It was the finest embalming that I have ever seen.
> On the right side of the throne sat a high priest dressed in a gorgeous robe of gold thread with a yellow Roman helmet-like hat. In front were two double rows of seated lamas facing each other. Between the pauses of the high priest's prayers the voices of the seated lamas swelled into a barbaric chant broken by the clash of cymbals. The air was heavy with burning incense.
> Mac and I kneeled with the others, touching our foreheads to the floor between our outstretched palms. . . . We stayed for half an hour doubled up like jackknives until I thought my neck would break. . . . We joined a group which drifted out of the temple into the courtyard. Then each of us had to whirl a dozen prayer wheels, murmuring the Tibetan invocation *Om mani padne om,* "The Jewel of the Lotus." Thousands of Mongols were streaming into the temple as we left, for the ceremonies continued all day long.

When Andrews' car drove away from Urga on the afternoon of June 6, he and Young were relieved to escape the treachery of the city for the serenity of the desert. The two government agents bounced along with Tserin on the piles of baggage, mail, and a large box of phonograph records sent by Yvette. One of the new passengers was Dalai Badmajapoff, the explorers' coconspirator at the Hutukhtu's funeral. The other man, John Dimschikoff, could only be regarded as a loathsome scoundrel who was already primed to send his superiors outrageously false and damaging reports. Although only twenty-four years old, he proclaimed himself a professor at a school in Urga, where he taught "every known branch of science." Later, at Tsagan Nor, the White Lake, the expedition was unexpectedly joined by a third member of the Secret Service, who, like Dalai, turned out to be honest and straightforward.

Evidence of how rapidly the Communists were tightening their grip on Mongolia was no longer limited to Urga, now known as Ulan Bator, "Red Hero." Andrews observed that official *yamens* were beginning to appear in extremely remote sections of the countryside. All caravans were inspected and taxed and, just as odious, the nomads were gradually being subjected to unreasonable restrictions. No one could sell their livestock without reporting the details. Taxes were levied on all such transactions, and nomads were forbidden to leave their circumscribed localities without permission—regulations incompatible with the free-spirited desert dwellers so admired by Andrews but which marked the instigation of a nationwide program of collectivization.

Andrews and Young returned to the Flaming Cliffs after a two-week absence. Granger had sent the caravan westward to await them at Tsagan Nor while the scientists continued working at Shabarakh Usu, which was still yielding important fossils. Among the mail Andrews brought from Urga was a letter to Granger from W. D. Matthew at the Museum. As Andrews considered Matthew "one of the least excitable men I know," it was uncharacteristic that his letter overflowed with enthusiasm regarding a seemingly insignificant skull found by Granger at the Flaming Cliffs in 1923. Slightly over an inch long and encased in a sandstone concretion, Granger, with no opportunity to study it in the field, had labeled the skull as that of "an unidentified reptile." After examining the specimen in the Museum's laboratory, Matthew realized that it was actually the skull of a mammal from the Cretaceous period, an animal that had existed alongside the last of the Gobi's dinosaurs. Because the implications of this revelation with regard to mammalian evolution—and Osborn's theory in particular—were momentous, Matthew had concluded his letter to Granger: "Do your utmost to get some other skulls."

Amazingly, within an hour, Granger returned to camp with another tiny skull, almost identical to the first, embedded in one of the thousands of small sandstone concretions that littered the Flaming Cliffs, making the search for small fossils that might be concealed inside exceedingly tedious. Yet Granger and his assistants eventually found additional skulls, and some of the concretions contained bits of skeletal fragments. Within the next few weeks, Andrews reported, "we had obtained eight partial skulls with associated lower jaws, one skull without jaws, a fragment of a maxilla and parts of a mandible; the remains of eleven individuals in all. These have been assigned to five genera and six species." Ranging in size from shrews to rats, they represented incipient marsupials, insectivores, and one offshoot known as multituberculates—rodentlike creatures whose evolutionary fortunes ended about 15 million years into the Cenozoic period, or Age of Mammals, which dawned approximately 65 million years ago. Only the marsupials and insectivores survived, eventually giving rise to the genesis of the earth's extraordinary array of mammals, including mankind.

With typical hyperbole, Andrews proclaimed these unimpressive-looking mammalian remains, which were wrapped in cotton and placed in a cigar box, as "the most important discovery of the expedition," perhaps partly out of deference to Osborn's hypothesis involving mammalian origins that had launched the expeditions. Yet their extraordinary scientific value was inescapable. On November 20, 1925, Andrews, who carried the cigar box to New York in his suitcase, had the satisfaction of presenting the precious fossils to Matthew at the Museum.

When their work at the Flaming Cliffs was finished, the explorers moved on to Tsagan Nor, where the expedition had camped in 1922 when the lake was several miles across. Now they found it reduced by drought to a quarter of its former size. Wide expanses of foul-smelling

mud encircled the shore, and the lush green vegetation along its edge
had turned to dull yellow grass. The newcomers to the expedition regarded with skepticism the veteran explorers' glowing descriptions of a
jewel-like lake set amid sand dunes with the majestic peaks of Baga
Bogdo rising in the distance. The often acerbic Nelson commented
tersely, "Tsagan Nor. It's little and it stinks." An aged nomad who had
lived his entire seventy-three years in the area told the explorers that the
lake disappeared twice in his lifetime. By the following year it had completely dried up again.

From Tsagan Nor, Granger, Olsen, and Liu moved fifteen miles
north to the Loh badlands, where a *Baluchitherium* skull had been recovered in 1922. Sticking out of a slope of eroded sediments, Liu spotted the glint of white bone. When he and Granger probed deeper, they
uncovered a gigantic lower-leg bone and foot of a *Baluchitherium*
standing upright, "just as if the animal," wrote Andrews, "had carelessly
left it behind when he took another stride." Fossils are so seldom found
in this position that Granger, momentarily baffled, sat beside the bone
for a long time contemplating possible explanations. "Quicksand!" he
suddenly exclaimed. "It was the right hind limb that Liu had found;
therefore the right front leg must be farther down the slope."

Measuring off about nine feet from the hind foot, they began to dig.
Just as Granger had suspected, another leg bone was located in the
same upright position. Soon the two other limbs on the opposite side
were exposed. All four stumps were situated exactly where they should
have been, each one in its separate pit. "The effect was extraordinary,"
Andrews recalled. Probably the titanic beast had come to drink from a
pool of water covering treacherous quicksand. Suddenly, it began to
sink, settling slightly back on its haunches, struggling to free its enormous body until the sediments suffocated it. Over the years, the rest of
the skeleton had weathered away, leaving only the lower leg bones cemented in the hardened earth. "The sight of these four stumps,"
Granger wrote, "and the frantic struggle they represented, allowed us to
envision a remarkable paleontological drama."

While the scientists continued working around Tsagan Nor, Andrews decided to reconnoiter the territory farther to the south and west,
even though he had learned from nomads that the terrain was a morass

of obstacles. Accompanied by Loucks, Shackelford, and Young, he struck south to Ikhe Bogdo (the Great Buddha), a range of jagged peaks west of Baga Bogdo. At its base lay a glistening lake, Orok Nor, which had attracted flocks of water birds and bats.

Unable to reach Orok Nor because of sand dunes blocking its approaches, the car struggled to the summit of a ridge of mountains from which Ikhe Bogdo was visible in the distance, although the ruggedness of the land made further travel impossible. On the return trip, they came to another sizable lake known as Kholobolchi Nor. Encircled in places by lush green grass, it was so inviting that Andrews moved the base camp to its shores on June 28.

Andrews next sought to penetrate beyond the Altai Mountains. Again traveling with Young, Shackelford, and Loucks, he drove in the direction of the towering Baga Bogdo, hoping to traverse "the massive ramparts," as Andrews described the Altai, "banning us from the south." They found no usable trails, only canyons, dry streambeds, alluvial fans, rocky outcroppings, and awesome gorges flanked by rugged mountains — "passable for horses and camels without a doubt," Andrews observed, "but hopeless for cars."

On July 1, Granger, Lovell, Berkey, and Andrews again set out from Kholobolchi Nor to explore the region as far west as the longitude of the trading center of Uliassutai. It proved to be a difficult and unrewarding venture over hazardous terrain. Utterly barren, the only signs of human life were a few impoverished nomads and a partially decayed corpse sprawled in the sand. With gas running low and no indications of new fossil deposits, the explorers turned back. Altogether, Andrews' reconnaissance of these southern and western regions had covered approximately twenty-two hundred miles.

Spurred by rumors of war in China, increasing hostility of the Buriats encountered at *yamens*, and the approach of cool weather, the expedition reassembled at Shabarakh Usu for the return to Kalgan — planning to stop again at Ula Usu so Granger could complete paleontological investigations begun earlier that summer. On the morning of August 2, the explorers left the Flaming Cliffs. "This single spot," Andrews reflected nostalgically, "had given us more than we dared to hope for from the entire Gobi. . . . I was filled with regret as I looked for the

Embolotherium

A type of titanothere, Embolotherium ("the battering ram beast," named by Henry Fairfield Osborn) probably used the curious nasal projection located in front of its head primarily for defending itself from predators.

last time at the Flaming Cliffs, gorgeous in the morning sunshine of the brilliant August day. I suppose I shall never see them again!"

When the digging at Ula Usu was completed, a camp was established on a rim above the lamasery of Baron Sog. Here Olsen excavated a huge titanothere skull discovered by Loucks and later named *Embolotherium loucksi*—a remarkable-looking beast with a snout shaped like a battering ram. Ten miles farther north, at the next camp on the trail to Kalgan, Andrews was horrified to discover three pit vipers of the genus *Agkistrodon* close to the tents. Brown, thick-bodied, and quite venomous, they were similar in appearance to the North American copperhead. Although the explorers had encountered a few snakes elsewhere in the desert, they tended to ignore them. But a rapid drop in temperature suddenly brought an invasion of the reptiles slithering into the new camp in search of warmth, causing everyone, Andrews said, "a very busy night." Within minutes, the tents were alive with snakes. Lovell saw one writhing through his doorway, but before leaping out of bed to kill it, he wisely thought to check the ground with a flashlight. Wrapped around all four legs of his cot were vipers, which he methodically butchered with a geologist's pick; another crawled out from under a gasoline tin near his head. Lovell disposed of several more, while Morris killed five in his tent. Wang, the Chinese chauffeur, found one coiled in his shoe and another fell out of his hat. Soon the Chinese and Mongols dashed for the safety of the automobiles to spend the night. Armed with a hatchet, Andrews attacked something soft and round under his foot only to discover it was a piece of rope. Moments later, Granger lunged viciously at what turned out to be a pipe cleaner, which he sliced neatly in half. Forty-seven vipers were killed that night. Luckily, the cold weather had made them sluggish and slow to strike. Only Andrews' German shepherd, Wolf, the expedition's mascot, was struck in the leg, but he was treated by Olsen and quickly recovered. "The new camp proved to be just as rich in fossils as in snakes," Andrews noted, "but at last the snakes won. Moreover, flurries of snow warned us to be on our way southward. On September 12, the cars roared down the slope to the basin floor, leaving [what we called] Viper Camp to the snakes and vultures."

Andrews left Outer Mongolia just in time to get his men and equipment out of the country before a firestorm erupted. It actually began on August 15, as the result of an article in the *Ulan Bator—Khoto News* attributed to the overzealous Communist John Dimschikoff, one of the government agents who had accompanied the expedition. Andrews knew nothing of the situation until he returned to Peking on August 20 for a four-day business trip. To his dismay, the city was buzzing with press reports that his expedition had been expelled from Mongolia. Accusing him of violating his agreements, the government had charged Andrews with spreading anti-Bolshevik propaganda, searching for oil and minerals, plotting with foreign powers to take over the country, and employing "a number of persons held suspicious in a military sense."

Dimschikoff had plied his superiors with enough venom regarding Andrews to exceed their worst suspicions. His blast in the *Ulan Bator—Khoto News* related how the expedition was guilty of mapping restricted areas; the "great Andrews," Dimschikoff alleged, allowed no one but himself to hunt animals, arrogantly ignored the Mongolian observers, refused to let them speak except in whispers, regarded them as a "lower race," and confined them to their tents. On the Fourth of July, the report continued, the explorers drank so much whiskey they threw each other into a lake in their evening clothes (formal attire in the Gobi!), and some of the men lost their shoes and watches in the water but were too drunk to look for them. With tales such as these being recklessly circulated, it was no wonder that the Ulan Bator paper demanded in an angry headline: ANDREWS EXPEDITION MUST LEAVE MONGOLIA.

Matters became even more complicated when Dalai Badmajapoff and the Mongolian government's third agent, who had joined the explorers at Tsagan Nor, returned to Ulan Bator and emphatically refuted Dimschikoff's allegations. But Badmajapoff's unequivocal defense of Andrews and the expedition's integrity triggered a mysterious incident that has never been explained. Andrews later heard from reliable sources that the notorious Dimschikoff was banished to his native Siberia for misrepresenting the scientists' activities. Two years later, he returned to Ulan Bator and was dispatched, again with Dalai Badmajapoff, to Germany on official business. Supposedly, Dimschikoff stole their expense money and poisoned Dalai, claiming that Dalai had absconded with the funds and committed suicide when Dimschikoff unveiled the theft.

Infuriated by reports of his expulsion from Outer Mongolia, Andrews and the Museum responded vigorously to the media:

> It is hardly necessary to deny the reported charges of political propaganda and military mapping. The [Expedition] had written permission from the Mongolian Govt. for the itinerary and locality maps which we made as well as for all the other investigations which it carried on.
>
> It would be the height of folly for any scientific expedition to concern itself with the politics of the countries in which it conducts its explorations. The [Central Asiatic Expedition] never has, and never will do so. We are explorers, not politicians!

As a result, tensions were somewhat eased when the Scientific Committee in Ulan Bator, embarrassed by international reaction to the Dimschikoff scandal, sent the following cable to the Museum:

> Our government had no intention of expelling the Expedition from the territory of the Mongolian Republic and informa-

tion in the press regarding this is wholly false; therefore, we express hope that you will according to your agreement with the Scientific Committee send everything promised and we also hope that you will personally come to clear up the above questions and establish yet more close friendship.

[Signed] President Scientific Committee Jamyan,
Perpetual Secretary Jamtsaraw

After his previous encounters with Communist officials, Andrews showed no enthusiasm for returning to Mongolia under the illusion of achieving a "close friendship." And it was not until 1930 that the last of the duplicate scientific material arrived at the State Museum in Ulan Bator—fourteen crates sent by the American Museum containing fossils, casts, and photographs. Andrews' position had also been compromised by ambiguities in his statements to the press. If indeed he believed it was the "height of folly," as he asserted, for a scientific expedition to concern itself with the politics of a host country, his robust denouncements of the Communist takeover of Outer Mongolia belied such sentiments. Andrews loathed the Bolsheviks, and made no effort—privately or in public—to disguise his attitude, especially his intense hatred of Communist officials who tried to impede his explorations.

Nor can one completely dismiss the Mongols' concern that the expeditions were searching for minerals and oil. Given Andrews' Wall Street backers and corporate support (Standard Oil supplied four thousand gallons of gasoline and one hundred gallons of oil for the 1925 season alone), it was a legitimate issue. While there is not a shred of verifiable evidence that Andrews' expeditions ever engaged in such activities, the American Museum's archives contain several letters from Standard Oil executives asking Andrews to report any promising oil-bearing formations the geologists might discover. One suspects that Andrews perfunctorily agreed as a courtesy to Standard Oil for its assistance, then dismissed the whole business. Any prospecting for oil or minerals, as Andrews repeatedly acknowledged, would have violated his permits with the Mongolian government, ended the expeditions, and discredited his reputation.

Military activities, however, were another matter, though clearly a harmless one without sinister implications. It was well known that Andrews submitted classified reports on a variety of subjects—first to naval intelligence during World War I and later to the War Department. Frederick Butler, on loan to the 1925 expedition from the Army Corps of Engineers, compiled a fifty-page report on his observations, which was intended to familiarize his commanding officers with a wide range of information pertaining to Mongolia's economy, geography, transportation, population, military capability, and climate. Although it contained nothing that was not common knowledge to most travelers in the country, copies of Butler's report were sent to the chief of army engineers in Tientsin and the adjutant general in Washington, where they soon disappeared into the vast storehouse of bureaucratic paperwork on Asia.*

Frederick Butler was a keen observer of people and possessed a wicked sense of humor. In his journal, he recorded incisive vignettes of his companions. Granger, for example, "looked more like Buffalo Bill than a Ph.D."; Shackelford was the "court jester"; Berkey was "the patriarch"; and Young's life "was a romance in itself." Andrews, he observed, had a "very nervous and restless disposition . . . whose life will make classic reading if it is ever compiled." "Our scientists are a queer assortment," Butler concluded. "They are all good sports, full of humor, and indeed inclined to be a bit on the shady side in their stories." With two or three minor incidents, "we all got along exceptionally well . . . [and] when it was over we seemed to part friends." In view of the difficulties, Butler felt the expedition was masterfully conceived and executed. As always, there was nothing penurious about Andrews' approach to life in the field. With his flair for style and comfort, each member of the staff wore custom-knitted sweaters emblazoned with the expedition logo, dined whenever possible in a large blue mess tent decorated with Chinese good-luck symbols in gold, and sat at tables cov-

*Intelligence gathering by the Museum's staff was regularly practiced on expeditions into remote or militarily sensitive areas well into the 1920s. A letter from the director, Frederick Lucas, to the War Department in Washington dated March 24, 1920, was one of several offering the assistance of scientists in the field (including Andrews, whose spying activities for naval intelligence were known to Lucas). "We shall always be glad," he wrote on the Museum's behalf to the War Department, "to do anything to be of service."

ered with white tablecloths, a luxury Andrews insisted upon. "Our food was excellent," Butler commented, "and in sufficient variety to keep our digestion and disposition in order."

Once back in Peking in late September, Berkey, Morris, and Roberts returned to the United States. Shackelford followed soon afterward with twenty thousand feet of film he had processed at the headquarters. Dr. Loucks resumed his position at the Peking Union Medical College, and Lieutenants Robinson and Butler rejoined their military units. George Olsen, assisted by five Chinese preparators, expanded the paleontological laboratory at the headquarters to enable them to clean and restore fossils as they were uncrated from the field.

Ralph Chaney went northward into Manchuria, where he examined large quantities of fossilized vegetation near Mukden. Buried in shales and coal, he found remnants of what had once been thick forests of redwood, oak, alder, maple, sycamore, poplar, and ferns. In the meantime, Granger returned again to the fossil pits at Yen-ching-kou in Szechwan to resume his study of their seemingly inexhaustible Cenozoic fauna.

On December 3, Nelson set out from the port of Ichang to explore caves along the banks of the Yangtze River in hopes of finding evidence of prehistoric man. Accompanied by his wife, a Number One Boy, an interpreter, and a cook, Nelson conducted most of his research from a sixty-eight-foot junk manned by its owner and a crew who maintained an opium den in a forward compartment. Over a four-month period, they navigated the Yangtze for a distance of approximately 230 miles, investigated 367 caves and shelters, and noted the location of 316 additional sites. But except for a handful of Neolithic artifacts, the results of all this effort were extremely disappointing. "From the start," Nelson observed, "nothing really worth the name *prehistoric* turned up and before long it became tolerably clear that nothing much was to be expected."

With these varied activities under way, Andrews had sailed for the United States to engage in more fund-raising and another lecture tour. Mac Young, who by now practically ran the logistical aspects of the expeditions, assisted by Lovell, settled into the headquarters, where Yvette, George, and Kevin were also comfortably ensconced, to prepare for the 1926 expedition that was to concentrate on Inner Mongolia. As usual, everybody was nonchalant regarding the looming shadow of war in northern China. No one suspected that the long-burning fuse was about to ignite an explosion with terrible consequences that would impair the future of the Central Asiatic Expeditions.

CHAPTER 15

While Andrews was steaming toward America, the ship's radio picked up the news that civil war had erupted near Shanghai between the forces of Feng Yu-hsiang and his archrival, the former Manchurian bandit Chang Tso-lin. At first, Andrews was not unduly worried. Shanghai was far removed from Peking, and he felt the war would not spread northward. Even if it did, generals on one side or another would probably be bought off, thus ending the hostilities. At least that was a common scenario in past conflagrations.

But this time Andrews was mistaken. Some of the bitterest conflicts took place as far north as Tientsin, with terrible casualties. Railroads throughout northern China were paralyzed, making food in Peking scarce and expensive. Atrocities were commonplace, especially public beheadings, torture, and firing squads. Shops in Tientsin were barricaded, and the streets were nearly deserted.

Given these conditions, Mac Young faced enormous difficulties trying to organize the 1926 expedition. The huge amounts of flour, rice, coffee, sugar, beans, and other necessary staples could only be purchased in the port city of Tientsin, and 3,500 gallons of gasoline, again donated by Standard Oil, were stored there in warehouses. Without rail service, there was no way to transport anything from Tientsin except by car, and for three weeks, Norman Lovell made a round-trip daily during a lull in the fighting. Leaving Tientsin early every morning in a heavily loaded car, he drove the eighty miles to Peking by noon, ate lunch at the headquarters while servants unloaded the car, and started back to Tientsin around one o'clock. He was repeatedly stopped and questioned by soldiers, but a plentiful supply of cigarettes got him safely through the lines.

Not long before Andrews returned to China in late March 1926 — having raised approximately $50,000 during his lectures in the United States — four Dodge Brothers cars, specially ordered for the expedition, arrived at Tientsin from Detroit. Lovell and three Chinese chauffeurs arranged to drive them to Peking, but so many unruly soldiers piled into the empty cars they nearly destroyed the heavy-duty springs and ruined the gears.

By early March, the equipment and vehicles had assembled at the headquarters compound. The caravan should have already started for its first rendezvous. Instead, Young instructed Merin to conceal the camels in the desert for safety when Feng's rampaging troops began confiscating every horse, camel, mule, and cart within a hundred-mile radius of Kalgan in preparation for an assault on Peking. All cars belonging to the Peking-Suiyuan Railroad had been commandeered and it was impossible to move freight between Peking and Kalgan, even by offering to pay exorbitant squeeze. Moreover, all roads into Kalgan were heavily guarded by troops with orders to allow no one to pass.

On March 27, Andrews docked at Tientsin, accompanied by Shackelford. Young drove over to meet them despite bloody skirmishes close to Peking. Leaving Tientsin two days later, they were warned that Chang Tso-lin had mined the road in fourteen places. Andrews, Young, and Shackelford decided to chance it anyway. Either the car was not heavy enough to detonate mines designed for tanks and trucks loaded with troops, or luck was on their side, but they miraculously avoided being blown to bits. A week later, another stroke of good fortune came Andrews' way when his friend C. S. Liu, director general of railroads, managed to secure two freight cars to haul the expedition's supplies to Kalgan. On April 13, under a veil of secrecy, Merin brought the camels from their hiding place in the desert to be loaded. Young had come from Peking by train to assist Merin after securing a special permit from the local commandant that would allow the caravan to depart and await the cars at a well in the valley of the Shara Murun River.

Meanwhile, the staff began assembling in Peking. Shackelford had returned with Andrews, and since Lovell had previous commitments, Young was placed solely in charge of motor transport. Nels Nelson was back from his disappointing search for prehistoric man along the Yangtze River; Granger arrived after his third winter at the fossil pits in

Szechwan Province. Throughout 1925 and 1926, the herpetologist Clifford Pope had continued adding to his extraordinary collection of reptiles, amphibians, and fish in Fukien and Kiangsi provinces. In addition, three newcomers had joined the expedition's staff. Berkey, unable to take more time away from his other professional commitments, sent over one of his protégés as the geologist, a British graduate of Columbia and a Rhodes Scholar named Radcliffe Beckwith. Captain W. P. T. Hill of the Marine Corps was engaged as cartographer; and after several years of planning, William Diller Matthew came out from the Museum to join his close friend Walter Granger in handling the paleontology.

As a precaution, Peking was hastily sealed off against any attempted onslaught by Feng's army. The gates were blockaded and heavily guarded. Every Chinese with political influence had sought protection inside the legation quarter. Many foreigners who owned cottages at the racecourse, seven miles west of town, were isolated by bands of roving soldiers until they could be rescued. Notices had been distributed to all foreign residents instructing them where to assemble in the event of an emergency. Andrews had arranged for Yvette, George, and Kevin to take refuge in the American legation, while he devised plans to protect the compound that included posting sandbagged machine guns on the roof, positioned to sweep the surrounding *hutungs* from all angles. After a few days of unrelenting tension, "during which," Andrews wrote, ". . . a strange [aura] of preparation for a great calamity pervaded the air," it was learned that Feng had withdrawn his army, retreated along the railroad, and occupied Nankow Pass, the principal gateway in the Great Wall on the way to Kalgan. With the simultaneous advance of Chang Tso-lin's forces toward Peking, Feng had decided to reconsider his strategy.

On the morning of April 14, Andrews, Beckwith, and Hill tried to reach Tientsin to pick up supplies. For the first few miles, they drove unmolested through large contingents of cavalry and infantry. As they approached an ancient marble bridge at Tungchow, fourteen miles from Peking and under Chang Tso-lin's control, a bullet suddenly struck inches from the car's front wheel. Within moments, bullets were splattering around the car "like hailstones," Andrews recalled, although the reason for the attack was unclear. Suddenly, a group of soldiers opened fire with a machine gun, "but it was aimed too low and bullets were kicking up dust just in front of us. [They] could see the American flag [on the car] plainly enough but that made not the slightest difference."

Andrews whirled the car around and hastily retreated. For three miles, they ran a gauntlet of rifle fire from both sides of the road. Just as things were beginning to calm down, more trouble arose when three infantrymen asked for a ride. Thinking they might offer a measure of protection, Andrews let them stand on the running board. Almost immediately, one of the men slipped backward and fell to the ground. His hand slid under the wheel and was badly mangled, requiring a tourniquet to slow the bleeding. Ignoring Andrews' protest, more soldiers piled onto the car farther down the road. Inevitably, another accident occurred when one of them fell off and suffered a broken leg. Enraged, the soldiers stopped the car and ordered Andrews, Shackelford, Hill, and Beckwith to get out and line up. Raising their rifles, they had every intention of killing the Americans on the spot. Seconds before the soldiers opened fire, an officer appeared who could speak Mandarin (the infantrymen spoke only a difficult Shantung dialect). When Andrews explained what had happened, the officer directed them off the road to a barely passable route through cultivated fields that eventually brought the explorers to Peking's gates.

About this time, Peking experienced a new twist in its on-again, off-again wars. Every morning at ten o'clock, a single airplane began flying over the city. It would drop eight or ten bombs in the vicinity of the Chien Men railroad station and then disappear. Although the rail yards were the pilot's primary target, the plane, a World War I contraption sent by Chang Tso-lin, flew so high the bombs often went astray. Civilians were sometimes killed or injured by the small black-powder bombs, one of which hit a school, resulting in the death of several students.

Always eager for an exciting new distraction, the members of the foreign colony inaugurated a diversion known as "bombing breakfasts." Every morning, someone would invite a group of guests for breakfast at nine o'clock at the Peking Hotel. At 9:55, they all adjourned to the rooftop, which overlooked the Chien Men station. With drinks in hand, they would watch the airplane's antics, observe the explosions, then climb in their cars and rush off to inspect the damage.

One morning, Andrews gave his own bombing breakfast, but was disgusted when the plane failed to show up on schedule. At eleven, he had an appointment at the train station to discuss the possibility of shipping his automobiles to Kalgan. Just as he pulled up to the depot with Lo, his Number One Boy, the overdue plane appeared and began dropping bombs on the rail yard. Two bombs barely missed Andrews' Dodge before he and Lo scrambled under a freight car with a dozen terrified Chinese. After a few minutes, thinking the raid was over, Andrews and Lo crawled out. Suddenly, the plane reappeared and began pounding the station with more bombs, one of which exploded directly in front of a Chinese woman. "It blew her head off," Andrews remembered vividly, "as neatly as though it had been severed with a knife." After a few weeks, the air raids stopped as abruptly as they had begun, without one train ever having been hit.

On April 15, Andrews and Shackelford decided to join Young in Kalgan to help him get the caravan under way, but they were unable to get through since the road to Nankow was barricaded. Andrews then tried to reach Kalgan by going west into Shansi Province; that, too, proved impossible because of raiding and looting by Feng's troops. Unable to contact Young by cable, telephone, or through the American consul in Kalgan, Andrews grew alarmed. Not until two weeks later did Young turn up in Peking after a harrowing trip during which he walked several hundred miles across Shansi and lived on little else but salted eggs. Weak, unshaven, and hollow-eyed, he required several days of bed rest before regaining his strength. He brought word that the caravan was safely in the desert, but confirmed Andrews' opinion that there was no hope of getting the expedition to Kalgan through Shansi, as the countryside swarmed with too many brigands and guerrillas.

Andrews next set out with Lovell to see if Kalgan was accessible via Jehol, a city 140 miles from Peking that was once the summer residence of Manchu emperors. It was controlled by a military governor, a former bandit chieftain, who received Andrews and Lovell cordially and served them tea in a room where numerous automatic pistols hung from the walls and graced his desk. Allowing the explorers to pass beyond Jehol, the governor stated, would be signing their death warrants. The region was a hotbed of brigands and unpaid defectors from Feng's army who would stop at nothing to acquire the expedition's cars, gasoline, and supplies.

By July, Andrews reluctantly announced that the 1926 expedition would have to be abandoned. Matthew, profoundly disappointed at not being able to work with Granger in the Gobi, returned to New York by way of India. Shackelford booked passage across the Pacific, and Captain Hill was assigned to the American legation guard. Beckwith settled down in Peking to study Chinese and Russian during the winter. Granger and Nelson mapped out a paleontological and archaeological reconnaissance of Yunnan Province based on Andrews' previous explorations there in 1916–1917. Since no news had reached Merin after his arrival at Shara Murun, he headed back toward Kalgan, trying to avoid Feng's soldiers. Believing Merin was lost or taken captive when he was not heard from for weeks, Young organized a search party. But the caravan finally made its way to a Swedish mission at Hallong Osso, which was operated by Andrews' friend Joel Eriksson. By now, only eighteen foreigners still remained in Kalgan. Food was desperately scarce, and in response to an urgent request from the American consul, Andrews sold him the supplies from his caravan at wholesale prices to help ease the shortage.

As there was nothing Andrews could accomplish in Peking, he sailed from Shanghai on September 1 for another fund-raising tour of the United States, leaving Mac Young in charge of expedition affairs. Yvette and the children had moved back into the headquarters compound, where they were under the protection of Young, the servants, and a small detachment of legation guards. "Our fruitless summer," Andrews wrote, "had cost considerably more than would have been expended during a season's field work, and more money was urgently needed."

In October, he sailed from New York for England on the *Aquitania*, where on November 8 he delivered the Second Asia Lecture before the Royal Geographical Society in London. Andrews' talk, illustrated with color plates, was an overview of the achievements of the Central Asiatic Expeditions' 1922, 1923, and 1925 seasons. He drew a packed house.

On January 26, 1927, Andrews interrupted his lecturing and fund-raising activities to enter New York's Presbyterian Hospital for surgery to

RIGHT: Reconstruction drawing of *Andrewsarchus* from Irdin Manha. Drawn by E. R. Fulda.

LEFT: *Entelodon,* a giant piglike mammal that roamed widely over the Gobi. From a painting by Charles R. Knight.

RIGHT: Restoration of *Baluchitherium,* also known variously as *Paraceratherium* and *Indricotherium,* as envisioned by Henry Fairfield Osborn. Paleontologists have long debated details of its size, anatomy, and relationship to similar species. It has been argued, for example, that the short neck depicted by Osborn could have made it difficult for the huge animal to drink from pools or streams. Drawing by Charles R. Knight.

Nomads clad in typical riding boots with upturned toes and long robes, known as *dels*.

BELOW: Women of the Chahar district of Inner Mongolia in front of a yurt (a Turkic word), now known as a *ger*—the original name.

Watering camels at a desert well.

Interior of a *ger* showing typical furnishings and the collapsible, latticework frame covered by felt.

Norman Lovell repairing an engine.

The 1928 expedition passing beyond the Great Wall under Chinese military escort.

Staff of the 1928 expedition at Hatt-in-Sumu. *Seated, left to right:* Perez, Andrews, Granger, Spock, Thompson. *Standing:* Shackelford, Pond, Eriksson, Horvath, Young, Hill.

Andrews and Tserin at Hatt-in-Sumu, 1928.

ABOVE: Tserin and Andrews watch the supply caravan leaving Chimney Butte, Inner Mongolia, 1928.

ABOVE: Extricating a heavily loaded car. Often what appeared to be a solid gravel plain would give way to areas of mud or deep sand.

LEFT: Andrews and Hill surveying at Urtyn Obo.

Women of the Khalka tribe wearing traditional headdresses, Urga, ca. 1922.

The Gandan Lamasery, Mongolia's holiest temple, Urga, ca. 1922.

Nomad women wearing heavy silver cases, studded with turquoise and coral, that decorate their long braids.

RIGHT: A nomad encampment in the Chahar district of Inner Mongolia.

BELOW: Herdsmen carrying lassoes, formerly known as *urgas*, used to snare horses.

Andrews with a pet vulture.

Nomads watching Hill operate a short-wave radio.

Excavating the skeleton of a *Baluchitherium* at Holy Mesa, 1928.

Expedition visiting the Goptchil Lamasery, Inner Mongolia, 1928.

Andrews relaxing with a pet antelope.

Prospecting for fossils at Urtyn Obo, 1928.

Skeleton of an Oviraptor lying on a nest
of unhatched eggs (see Epilogue).

Pausing to cool the cars' engines en route from Inner Mongolia to Kalgan, 1928.

Excavating the lower teeth of a shovel-tusked mastodon or Platybelodon, Wolf Camp, 1930.

FIFTEEN CENTS

TIME

The Weekly News-Magazine

VOL. II NO. 9

ROY CHAPMAN ANDREWS
" Across the Gobi Desert"—
(See Page 18)

OCT. 29, 1923

Andrews on the cover of *Time*.

Andrews playing polo in Peking.

An execution carried out not far from Andrews' polo club. Another popular location for public executions was a field across the street from the Temple of Heaven in Peking. Rarely did such atrocities, war, or civil unrest disrupt the frenetic social activities pursued by members of the city's foreign colony.

Andrews and his second wife, Wilhelmina or "Billie," in their New York apartment, ca. 1936.

Living room of Andrews' apartment at 11 East Seventy-third Street showing a portion of his collection of Chinese art.

correct the injury to his collarbone, which he had dislocated in 1925 during the New Year's Day steeplechase in Tientsin. Because it had never healed properly, Andrews was suffering severe pain caused by a bone spur. Ten days after the operation, he was visited by his friend, the publisher George Palmer Putnam. For months, Putnam had been trying to induce Andrews to write a book on the first three seasons of the Central Asiatic Expeditions. Andrews had always begged off, citing his hectic schedule. Now that excuse was no longer valid, since he was laid up in the hospital with nothing to do but hold court with a steady stream of visitors and reporters. Putnam offered to send him a secretary to assist with the writing, and wrote a check for a $1,000 advance against royalties. Andrews agreed and started to work. When he moved to Osborn's town house for nearly two more weeks of convalescing, he continued to write in a room set aside for his use. By including articles he had published in *Asia* and dictating other chapters, the book was finished in a month. Titled *On the Trail of Ancient Man*, and with an introduction by Osborn, its popularity fully justified Putnam's enthusiasm for the project.

Once he had recovered, Andrews gave two lectures at Carnegie Hall before capacity audiences. Afterward, he embarked on a nationwide speaking tour amid a deluge of publicity. In February of 1926, he was awarded an honorary doctor of science degree by Brown University, which was followed in 1928 by a doctorate from his alma mater, Beloit College. Toward the end of 1927, Andrews had acquired a more prestigious title from the American Museum as well. Osborn named him curator of a newly created Division of Asiatic Exploration and Research. But the term *curator* was not to Andrews' liking. In a brash display of self-promotion, he convinced Osborn that it failed to convey the "overwhelming importance of the duties and responsibilities" he had assumed as organizer and leader of the Central Asiatic Expeditions. A more appropriate title, he argued successfully, elevated him to the position of Vice Director, Division of Asiatic Exploration and Research.

"Until early spring," Andrews wrote, "I did nothing but talk from one end of the country to the other. It netted fifty thousand dollars for the expedition. I sat at so many dinners and met so many interesting and important people that they formed a blur of places and personalities. Ending up on a Pacific liner, I slept most of the way across to

Japan." He was forced to enter China via Korea. With hordes of people fleeing the country, it was impossible to secure passage to Tientsin or Shanghai, though he eventually made it by train to Peking and a long-awaited reunion with Yvette and his sons.

The China to which Andrews returned was gripped by fear and violence. Student riots aimed at toppling the governing regime had broken out, encouraged by Bolshevik propagandists. And a new element had complicated the struggle between Chang Tso-lin and Feng Yu-hsiang with the emergence of a powerful army in the south commanded by Chiang Kai-shek, who was moving northward in yet another bid for control of the country. Antiforeign sentiment, which had long smoldered in China, had erupted in the so-called Nanking and Hankow "outrages," which occurred during the winter of 1926–1927. At Hankow, Chinese militants successfully seized the British concession with virtually no resistance. In Nanking, murderous gangs of Chinese soldiers killed many foreign residents, looted their homes, and occupied consulates. Forced to seek refuge on a hill, the foreigners were besieged by the rampaging soldiers until an American marine, at great personal risk, climbed to the roof of a building under heavy rifle fire and signaled a United States Navy gunboat. Its batteries laid down a barrage, scattering the Chinese long enough for a landing party to rescue the survivors. Additionally, the Chinese were demanding the return of the Tientsin and Shanghai concessions, and the abandonment of foreign legations in Peking, all of which had been granted to the United States and European powers as part of the treaty that settled the infamous Boxer Rebellion in 1900. Alarmed by the current situation, warships and thousands of foreign troops were sent to Shanghai and Tientsin to forestall threats of attack by Chinese militants.

When Andrews reached Peking, he found the foreign colony bordering on hysteria. "It was the first time," he wrote, "that I had seen anything like panic. Even the year before when the gates of Peking were closed and sandbagged, and Chang Tso-lin's wild Manchu hordes were looting and burning the countryside, few foreigners in the capital were even nervous. But the Hankow and Nanking outrages had awakened the anti-foreign feeling which exists in the hearts of most Chinese. The poorly veiled hostility [felt] . . . by all classes made us realize that wholesale slaughter had only been averted by the arrival of foreign troops."

Then an unexpected event drastically altered the atmosphere. Chang Tso-lin's troops carried out a daring raid on the Russian-owned Dalbank and the Soviet attaché's office in the legation district. Descending on the bank and the embassy in large numbers, Chang's soldiers confiscated documents proving that the Russians—relying on diplomatic immunity for protection—were using the Dalbank and the attaché's office as a clearinghouse for Bolshevik propaganda and a haven for secret agents throughout China, whose mission was to stir up unrest.

With the names of agents and the storehouse of covert information taken from the Dalbank's vaults, Chang Tso-lin, a rabid anti-Communist, ordered the roundup of all propagandists in north China. A number of agents caught in the bank were slowly strangled in Chang's presence. "Hardly a day passed," Andrews recalled, "that one or more persons were not executed at the public grounds opposite the Temple of Heaven." Once the propagandists began fleeing northern China en masse, Peking quickly settled down to a somewhat normal existence.

Nevertheless, sporadic fighting continued in the countryside around Peking. Inside the city, grim reminders of recent hostilities were still to be seen. Bamboo cages containing severed human heads dangled from trees and light posts. Decaying bodies occasionally sprawled in the streets or hung from branches. Men and women stripped of their clothing and horribly mutilated were often carried through the streets skewered on poles.

Andrews' son George vividly recalled a gruesome example of the disregard for foreigners, which had become increasingly common after 1925. Driving home with his father along a crowded street, he saw a man being dragged by soldiers from an apothecary shop. Accused of overcharging for candy, which the hapless vendor frantically denied, an officer ordered his execution. Looking for a suitable place to carry out the punishment, the commander instructed two of his men to force the victim to kneel and place his head on the fender of Andrews' car, which had stopped when the commotion spilled into the street. Despite Andrews' protest, an instant later a swordsman sliced off the man's head and impaled it on a bayonet. The hood, running board, and windshield of Andrews' Dodge—even the American flag on the fender—were splattered with blood.

In view of the political quagmire into which China was rapidly sinking, the prospects for another expedition in 1927 were untenable. Even had Andrews been inclined to risk it, the American minister, J. V. A. MacMurray, could not have allowed the expedition to proceed under such dangerous circumstances. Moreover, in March 1927, Mac Young suffered a serious case of frostbite that would have prevented him from accompanying Andrews into the desert. He had driven a hundred miles north of Kalgan to retrieve the camels in the event an expedition did materialize at the last minute. On his way back to Kalgan, a blizzard plunged the temperature to 40 degrees below zero, and by the time Young reached the city, his fingers were severely frozen. At Kalgan, he waited another night and day for a train, then endured a fourteen-hour trip to Peking. When he entered the hospital, his blackened fingers were rapidly turning gangrenous and the doctors urged amputation of all the fingers on both hands. Young flatly refused, saying he would rather die than lose the use of his hands. He consulted Harold Loucks, the expedition's surgeon in 1925, who offered to try an experimental and extremely painful procedure that—to everyone's astonishment—saved all but the tips of four fingers.

With the cancellation of the 1927 expedition, Andrews wrote to Osborn:

> I have all our valuables packed in a warehouse in the Legation Quarter, the cars are at the American Legation, and everything except 157 cases of gasoline, in Mongolia, [is] disposed of—I am negotiating for its sale. If looting occurs Young and I will remain here to protect the house.
>
> It is impossible, of course, to predict what events will take place. I do feel certain, however, that something . . . will occur this summer . . . which will determine whether or not we can continue working in [Inner] Mongolia. What I wonder about is what to do in case we have to give up work in Mongolia?
>
> I am not ready to settle down to a curatorial job yet. I have too much restless energy. I am too full of health and strength and enthusiasm for the field work in which I have made a success. I would be deathly unhappy to give up exploration.

_____

"Spending a winter in Peking can hardly be called one of the hardships of an explorer's life," Andrews once wrote. "It is the most interesting and . . . delightful city in the world and I say that advisedly. I lived in a beautiful old Manchu palace; had a staff of eighteen efficient servants; a stable of polo ponies and hunters; and a host of friends." Travelers from all over the world came to visit Andrews and Yvette, and were always entertained lavishly with dinners or receptions. Noël Coward was a particular favorite among the foreign colony. Douglas Fairbanks, Sr., whom Andrews met in 1924 while lecturing in Hollywood, was another popular visitor to Peking. Owen Lattimore, the journalist and Far Eastern expert, was a frequent guest at the headquarters, along with the celebrated explorer Sir Aurel Stein.

Outside the compound's high protective walls, Peking's age-old enchantment somehow endured in the midst of senseless violence and constant threats of war. Hundreds of pigeons with whistles tied on their tails still fluttered overhead. The streets were crowded with vendors, outdoor restaurants, puppet shows, magicians, acrobats, and fortune-tellers, many of whom played musical instruments to attract attention. Hawkers were selling kites gaudily painted to represent demons, butterflies, fish, birds, and dragons. Weaving among the crowds were rickshaws, bicycles, and men pushing wheelbarrows with strings of tiny bells woven into the spokes of their wheels. Old men smoking pipes sat for hours under trees tending songbirds and crickets in delicate bamboo cages, while others passed the time playing mah-jongg or checkers.

"Peking is the one place left in the world where one can live an Arabian Nights existence," Andrews observed. "One rubs the lamp and things happen. Don't inquire *how* they happen; just rub the lamp!" No better example of this reality could be found than giving a dinner party. "It was virtually impossible," he wrote, "to plan a dinner in Peking's cosmopolitan milieu for only one nationality." As he and Yvette looked over their table night after night, guests from five or six countries would be present, and once, out of fourteen people, nine nationalities were seated at the table. In the days before 1926–1928, when civil war seriously began to disrupt the country, Peking enjoyed the finest European wines, liquors, and beer to complement superb Continental cuisine. Andrews' cook had worked nineteen years in foreign legations. His mastery of French, Russian, Italian, and Swedish dishes was extraordinary, and each morning, he presented the day's menu written in French. Andrews insisted on Chinese food, which he relished, at least twice a week—as long as the cook avoided the so-called delicacies: chicken windpipes, fish stomachs, duck fat, camel humps, bear paws, duck tongues, and, worst of all, sea slugs. But the meatballs, chicken, fish, pork, cabbage, pheasants, quail, prawns, gazelle, wild boar, Peking duck, and eggs (even the famous "hundred-year-old eggs," which were actually seldom more than two years old, soaked in lime, and served with soybean sauce) were Andrews' favorites.

Once the final arrangements for a dinner had been decided upon with the cook and Number One Boy, the matter was "entirely out of your hands," Andrews wrote. "You say what you want and things happen as if you had actually rubbed that Aladdin's lamp. We often gave dinners for twenty-five or thirty people although our silver and glass are sufficient for half that many. I never ask where the rest is coming from but it always appears. I simply tell my boy who is dining that night and he borrows whatever is necessary from the servants of some of the guests. We have some particularly beautiful candlesticks and fish plates and I am sure to meet them wherever we dine in Peking if it is a big party." Everything was quietly worked out in advance, though arranging such details always involved extra squeeze among the servants.

Some of the most lavish dinners were costume events or planned around a theme. On one occasion, Andrews and an equally mischievous friend gave a dinner and invited only people who were known to be avoiding each other or not speaking because of petty feuds. Invitations were sent out by special "chit-coolies" or private messengers so that no one knew who else was coming. Andrews arranged to have the most ardent adversaries seated next to each other. After the initial shock had worn off, the dinner, enhanced by numerous bottles of champagne and endlessly flowing cocktails, began to melt the icy atmosphere as the guests realized they had been duped. The evening turned into a rousing success and some of the bitterest animosities were overcome as the result of Andrews' and his coconspirator's shameless gall. "I don't think all the quarrels stayed patched up," Andrews later remarked, "but I know some of them did." Not that it really mattered. He freely admitted that the evening had been conceived purely in the "spirit of devilment with no intentions of doing a good deed. That part of it came incidentally."

Any enclave as tightly knit, cosmopolitan, and isolated as Peking's foreign community, which numbered about two thousand permanent residents, was certain to be rife with gossip, romantic trysts, flirtations, and adulterous affairs. Hardly a day passed without fresh rumors and salacious details of the latest sexual intrigue or impending divorce making the rounds of the clubs and cocktail parties. Not even Andrews and Yvette, two of the colony's brightest lights, were immune to temptation and scandalous speculation.

Yvette, with her dark hair and eyes, slender, graceful figure, and aristocratic bearing, was one of the most admired foreigners in Peking. Friends had worried for some time that she and Andrews were drifting apart. Indeed, she made no secret of her plight to some of her closest confidantes. As was all too evident, Andrews had allowed the expeditions to override his obligations as a husband and father.

During his frequent and protracted journeys, either in the Gobi or on fund-raising tours in the United States, Yvette was often escorted to races, polo matches, dinners, and parties by a former English military officer, St. John Smallwood, a businessman and polo player who handled Andrews' insurance and investments. He and his wife had been part of Andrews' and Yvette's innermost social circle, though his wife returned to England in 1924 with their two sons. Andrews relied on Smallwood—known by the nickname "Chips"—to look after Yvette whenever he was away. It turned out to be an ill-advised request. Apparently, Chips was far too attentive, and he and Yvette fell into a passionate love affair.

Exactly when Yvette confronted Andrews with the reality of the situation and how much she revealed to him is uncertain, but in the summer of 1927, she asked for a separation. In a stoic letter to Osborn, Andrews confided, "It has definitely been decided for Yvette to take the children and go to England. It is the wisest course we both know. What the future will hold for us cannot be predicted now and time will tell. We have nothing to do but let time decide it. It is all perfectly clear to me. I've played the game according to the rules, and, after all, we can't do more!"

By August, Yvette and the two boys boarded the Trans-Siberian Railway bound for England. George was enrolled in the "Dragon School" in the town of Oxford, so-called because the school's crest bore the image of a dragon. Yvette took a cottage in nearby Islip, which also happened to be close to her lover's home. It was rumored that their relationship continued on and off during his periodic visits from China, which Chips left permanently in the 1930s to return to England. His wife probably knew of the affair, but being a devout Catholic, she never consented to a divorce. After four years at the "Dragon School," George Andrews returned to the United States, where he attended St. Paul's School in Concord, New Hampshire, and graduated from Princeton in 1940. Kevin, seven years younger, eventually entered the Stowe School in England, enrolled at St. Paul's, and later went to Harvard.

PART FOUR

Politics and Paleontology

[Andrews possessed] an insatiable appetite for kinesthetic experience. He was never merely content to see the world. He wanted to *feel* it—all of it—and wherever possible to wrench out whole chunks as scientifically priceless souvenirs.

—D. R. Barton

Near the end of January 1927, Andrews sent a letter to Shackelford, who was working in Hollywood as a cinematographer. He stated that China's political climate had exploded "in flames." Unless conditions improved drastically, all foreigners, he believed, "will have to clear out of the country in the near future." Despite the fact that Chang Tso-lin's raid on the Dalbank restored a semblance of stability to Peking, heavy fighting continued to rage elsewhere, and an unholy cadre of warlords was bidding in China's power struggle. One of the most opportunistic of these was the rapidly emerging Chiang Kai-shek, who would ultimately proclaim himself leader of the Nationalist Party or Kuomintang. With help from the Communists, whom they later reviled, the Nationalists had now become a force to be reckoned with in the chaotic state of Chinese affairs.

With the odds clearly mounting against him, Andrews' resolve only seemed to strengthen. On October 27, he wrote to Granger at the Museum, telling him, "Keep up the spirits of the Professor in regard to the Asiatic work, for I am very sanguine of success in getting away next year regardless of all the complications. . . . Once we are off in the desert . . . we [will] have no trouble."

While he waited out the political storm in Peking, reports flashed across the front pages of newspapers that further fueled Andrews' determination to return to the field. In 1926, the anatomist Davidson Black, of the Peking Union Medical College, and the geologist J. G. Andersson, of the Geological Survey of China, announced the discovery of two human teeth from a cave in a hill at Chou Kou Tien (now called Zhoukoudian), about twenty-five miles southwest of Peking. The appearance of another tooth in 1927 and a juvenile skull the following year prompted scientists to investigate the cave's limestone deposits. Excavations produced five more skulls, an assortment of arm and leg bones, and approximately 350 teeth belonging to perhaps forty individuals who had inhabited the cave around 420,000 years ago. (They were originally believed to be closer to 600,000 years old.) Although the now-famous residents of Chou Kou Tien were originally named *Sinanthropus pekinensis*, or "Peking Man," we know them today as *Homo erectus*—the same upright-walking creature as the celebrated Java Man found by Eugène Dubois in 1891. They represented one of the final stages in the evolutionary progression leading to modern humans, or *Homo sapiens*.

The effect of the discoveries at Chou Kou Tien on Osborn and Andrews was electrifying. If bones of hominids this ancient existed in China and Java, they speculated that fossils of much older precursors must lie buried elsewhere in eastern or central Asia. Osborn reasserted his long-cherished theory that Asia was surely the "Cradle of Mankind," and proclaimed his confidence that the transition from primates to hominids had occurred, if not in the Gobi, then somewhere around its periphery—proof of which, Osborn insisted, must now be sought with renewed vigor. "All that stands in our way is politics," Andrews railed impatiently, "which we must find a way to overcome."

In the meantime, however, thousands of miles from China, another historic discovery had occurred, which would figure prominently in the quest for mankind's ancestors. In 1924, the fossilized skull of a child, no larger than a man's fist, was unearthed in a limestone quarry at Taung, in the southwestern corner of South Africa's Transvaal. It came to the attention of Raymond A. Dart, an Australian paleoanthropologist in Johannesburg who, after careful analysis of the skull, named it *Australopithecus africanus*, meaning "southern ape from Africa." While the Taung skull exhibited essentially apelike traits, Dart was convinced that the teeth and braincase displayed unmistakable hominid characteristics that placed it at the very threshold of the human family. Dart's hypothesis, published in 1926, met with an outpouring of skepticism from the overwhelming majority of his colleagues, who either dismissed his conclusions outright or were unwilling to abandon their ethnically biased belief in Asia as mankind's homeland. Nonetheless, although Dart's Taung skull, now dated at around 2.5 million years old, received little credence from scientists for more than twenty years, it would ultimately play a major role in shifting attention away from Asia to Africa in the debate over human origins.

As far as Andrews and Osborn were concerned, the controversy surrounding the Taung skull was irrelevant at this point, since the oldest *confirmed* hominid remains—Java Man and Peking Man—had come from Asia. Andrews could think of nothing else but carrying the search for additional human fossils to Inner Mongolia. By now, though, obstacles other than China's civil wars were threatening his dream. Ever since 1925, antiforeign sentiment had been growing throughout China. No longer was it directed solely at colonial businesses and military installations. Scholars and explorers also were being targeted because of their activities, a reaction fostered in part by Andrews' ill-conceived dinosaur egg auction. Moreover, priceless stores of ancient manuscripts and works of art were being carried out of the country by archaeologist-explorers referred to by the Chinese as "foreign devils"—the likes of Aurel Stein, Sven Hedin, Langdon Warner, and Albert von Le Coq.

Such antiforeignism reached a crisis in 1927 when an attempt was made to block a large, well-financed expedition into central Asia organized by Andrews' close friend, the famed Swedish explorer Sven Hedin. On March 11, 1927, a headline in the *Peking Leader* boldly proclaimed: CHINESE SCIENTISTS ARE VERY MUCH OPPOSED TO FOREIGN EXPLORATIONS. The article reported:

> Holding that Chinese ancient relics and treasures should be explored only by the Chinese people themselves and that they should not be exported abroad to adorn museums in Europe and America, scientific organizations in Peking in a joint statement Tuesday announced formation of a united association to fight the efforts of various foreign scientific expeditions to search for remains of ancient man and other treasures of geological and archaeological interest in different parts of China.

At the heart of this nationalistic movement was a group calling itself the Society for the Preservation of Cultural Objects. As Andrews pointed out, the society "was an entirely unofficial organization, but by false publicity it had succeeded in arousing such popular indignation that the government authorities dared not ignore its activities." Andrews knew it was only a matter of time before this "Cultural Society," as the group was more commonly known, turned its wrath on the Central Asiatic Expeditions, a conspicuous example of the kind of "imperialist" activity the society abhorred.

Finally, Andrews' patience was exhausted. The civil war in the north had developed into a stalemate. It was evident that a showdown between Chang Tso-lin, Feng Yu-hsiang, and the new players on the scene would not take place until the summer of 1928 at the earliest. Andrews, therefore, made a daring decision: to slip into the desert during this hiatus with as little fanfare as possible. After cabling the Museum to send out the staff no later than April 14, Andrews, with the help of his missionary friend Joel Eriksson, who was acting as the expedition's agent, set about purchasing a new caravan. Since most of the camels in the region surrounding Kalgan has been confiscated by soldiers, this proved to be a difficult task.

Then there was the matter of bandits to be dealt with. Bloody fighting near Kalgan had reduced drastically the number of caravans entering the city, and virtually all business with Outer Mongolia was suspended for nearly two years. Hundreds of unpaid soldiers were forced to become brigands to survive, and they raided right up to Kalgan's walls. Automobiles and caravans approaching or leaving the city were attacked before they had gone fifty miles. Merchants were being ruined, and the brigands themselves were starving. The normally plentiful stores of food in the rich agricultural region at the edge of the Mongolian plateau had been pillaged.

Eventually, the Chamber of Commerce in Kalgan devised a plan to offset the problem. Local officials agreed that special "liaison bandits" could enter Kalgan and make private deals with caravan owners for the safe passage of their camels. Every caravan that paid $5 per camel was allowed to come and go unmolested, but the price for each automobile traveling to Ulan Bator and back was $100.

Andrews arranged to meet with one of the most powerful liaison officers in Kalgan to discuss the safety of the expedition's automobiles and caravan. "A few years before," wrote Andrews, "he had been a respectable landlord of one of the motor inns on the Urga trail. I knew him well and knew that now he was a head brigand. What is more, he knew that I knew it. But it would have made him 'lose face' to admit the fact. Therefore, we were introduced as though we had never seen each other." He was posing as a "general" with enough "soldiers" to guarantee the expedition's protection, at least for the first hundred miles beyond Kalgan, which were the most dangerous. After an hour of tea drinking and extraneous conversation—a customary formality—they got around to the business at hand. The bandit suggested the usual fee of $5 a camel. Andrews offered $1. Because the caravan transported almost nothing that brigands could use or sell, they eventually settled on half the going rate.

As for the $100 fee for each of the cars, Andrews deliberately adopted a vague stance. "We were uncertain when we would leave," Andrews said, "[or] how many cars there would be—I would get in touch with him later. But I did mention that after all we hardly needed protection for we would have thirty men in the cars, all would have rifles and, moreover, there would be a machine-gun which could shoot two hundred bullets a minute. I did not make a definite statement. That would have sounded too much like a threat. . . . I just murmured it as though I were considering the matter aloud. But he understood, and talk of 'protection' money for the cars was dropped forthwith." (Andrews' expeditions never actually carried a machine gun, though he occasionally alluded to such a weapon whenever he felt it might strengthen his position with bandits or hostile soldiers.)

Anxious to get the expedition into the field, Mac Young and Andrews began transporting equipment and supplies from Peking to Kalgan, despite having to pay outrageous duties. Taxes were levied on everything when it left Peking, again when it was halfway to Kalgan, a third time when it reached Kalgan, and once more as it left Kalgan. In addition, all along the way individual "agents" demanded generous amounts of squeeze. Altogether, Andrews complained, "it cost us about ten thousand dollars in taxes and squeeze to get our cars, camels, food and equipment out of Peking, and into and out of Kalgan."

In a shrewd diplomatic gambit, the American minister, J. V. A. MacMurray, called on the local dictator, Marshall Chang Tso-lin, and with delicate protocol informed him that the fate of the expedition rested in his hands. Flattered by this recognition of his power, Chang gave his consent for the explorers to work in Inner Mongolia and arranged for local authorities to issue the necessary permits. Even so, Andrews resolved to launch the expedition in secrecy, without newspaper publicity that would alert the Cultural Society's watchdogs to his movements. In the hope of avoiding the complications encountered by Sven Hedin, Andrews summoned all foreign correspondents and reporters for English-language newspapers in north China and persuaded them not to publicize the expedition until after its departure. Exactly as planned, the scientists were able to slip out of Kalgan and into the desert without a hint from the press.

On April 3, the rest of the staff arrived in Peking. Granger brought an expert collector from the Museum named Albert Thompson, who had many years of field experience to his credit. Another of Berkey's students, L. Erskine Spock, had been engaged as the geologist. Shackelford was back for his third season as photographer, delirious to be returning to the desert and away from Hollywood. A navy medical officer, Dr. J. A. Perez, joined the expedition as surgeon, and Captain W. P. T. Hill took a leave of absence from the legation guard to serve as cartographer. Assisting Mac Young with maintaining the cars was G. Horvath, an ideal choice, since he worked as a mechanic for Frazar, Federal and Company in Tientsin, agents for Dodge Brothers.

Nels Nelson, who never showed much enthusiasm for Mongolia or China, declined to return. Waiting in the wings, however, was a young archaeologist and Beloit College graduate, Alonzo Pond, a specialist in European prehistory who had led an expedition in 1925–1926 for the Logan Museum to North Africa to search for evidence of primitive man. When Nelson withdrew from the project, Andrews offered the position of archaeologist to Pond, who had come to idolize Andrews.

In need of work, the caravan leader, Merin, one of the stalwarts of Andrews' first three expeditions, had been engaged by Sven Hedin for his difficult trek into Chinese Turkestan. Andrews deeply regretted Merin's departure, as they had grown quite close, but his place was taken by the equally trustworthy Tserin, who had been with the expeditions since their beginning as an interpreter and all-around assistant. With the native personnel consisting of six technicians, three cooks, two camp boys, two Chinese chauffeurs, three Mongol interpreters, and twelve caravan drivers—plus the scientific staff—the expedition numbered thirty-seven individuals. In addition, there were eight Dodge automobiles, and a supply caravan of 125 camels.

Exhilarated to be back in the desert after two tension-ridden years, Andrews planned to proceed first to the region known as Shara Murun and resume study of previously discovered fossil beds. While this work was under way, a small party would strike westward, seeking a route into Chinese Turkestan. "Our particular objective," Andrews wrote, "was the great low basin which we had tried twice to reach in 1925 by crossing the Altai Mountains."

Relying on Andrews' agreement with the liaison bandit to afford protection, Tserin left Kalgan with the expedition's caravan early in March. He was instructed to drop extra gasoline at a well along the trail, then await the rest of the group at a lamasery in the Shara Murun valley, about two hundred miles farther into the desert. On April 12, the staff, except for Andrews and Hill, arrived at the British-American Tobacco Company in Kalgan. Two days later, the American minister, J. V. A. MacMurray, his wife, sister, and Lewis Clark of the American legation drove to Kalgan with Andrews and Hill to accompany the expedition as far as Changpeh.

On the morning of April 16, eight Dodge cars, flying the flags of the United States and the Explorers Club in New York, made their way through the Wanchuan Pass and the Great Wall. A ninth car transported MacMurray and his passengers. More as a gesture of protocol than anything else, the authorities in Kalgan provided an escort of fifty cavalry to accompany the cars through the pass and for a short distance toward Changpeh. Because of payoffs to liaison bandits and Chang Tso-lin's increased patrols, the danger was presumed to be slight, but officials at Kalgan were unwilling to risk the safety of the "American Envoy Extraordinary and Minister Plenipotentiary." At a point several miles beyond the pass, the cavalry halted to await MacMurray's return to Kalgan later that afternoon. The expedition proceeded to Changpeh, where it was to spend the night. After a farewell toast, MacMurray's party left the inn at Changpeh and, starting back sooner than expected, surprised the "gallant soldiers" in the act of robbing a caravan!

The officer in command of the garrison at Changpeh, who seemed trustworthy enough, warned Andrews that three hundred bandits were operating in the vicinity of Chap Ser, some sixty miles farther along the road. According to the commandant at Changpeh, his jurisdiction ended twenty miles ahead. Any armed men posing as soldiers beyond that point should be considered bandits.

Not far from a shallow stream called Black Water, eight well-armed riders in military uniforms appeared along the road. Instantly suspicious, Andrews ordered his companions to cover the horsemen with rifles, while slipping his revolver from its holster. Their spokesman announced that they had been sent by the commander at Changpeh to escort the explorers through a brigand-infested area. He invited them for tea at a village about ten miles ahead, which Andrews had been warned was actually a bandit stronghold. When Andrews asked to examine the soldiers' credentials, he was shown what was allegedly a letter from the commandant at Changpeh. Andrews noticed at once that it had no "chop" or seal, without which no written communications in China were valid. It was obviously a thinly disguised ruse, one that Andrews knew was commonly used by brigands along this stretch of the Kalgan–Ulan Bator road. Afraid to attack in the open, the bandits sought to lure their victims to a village under their control, where the explorers would have been murdered or held for ransom.

Andrews ordered the would-be assailants to leave under threat of being shot. Unwilling to challenge a phalanx of high-powered rifles aimed directly at them, the brigands panicked, whirled around, and galloped away. As they tore off, several of Andrews' men could not resist firing a few shots into the ground at their heels. When the cars passed the village a short time later, Andrews confirmed that it was a veritable bandit fortress, which they raced through at full throttle.

The expedition stopped briefly at one of Eriksson's missions at Hatt-in-Sumu before setting out on April 20 along the Sair Usu trail for the Shara Murun valley. Andrews was alarmed to see so many dead camels along the way, afraid they might be from his own caravan. But by late afternoon a white escarpment overlooking the Shara Murun valley burst into view. Soon Tserin and the camels were clearly visible. All were in good condition, though Tserin had stopped seven miles from the lamasery because of the excellent grazing in the region.

"We camped that night beside them," Andrews wrote. "For the first time I felt that the Expedition was really under way again after the two years' struggle with war and brigands, officials and diplomacy. The Gobi lay in front of us; our only opponents were the natural forces of the desert."

Pushing ahead past herds of gazelles, the cars slowly climbed to the edge of an escarpment. Following in the tracks of their 1925 route, they passed Viper Camp, with its unnerving recollections of the snakes that had invaded the explorers' tents. From here, the cars swung north to a place called Chimney Butte, where rich fossil deposits had been located on the previous expedition. A base camp was established on a ridge overlooking surrounding badlands on three sides—gray and red buttes, steep-sided canyons, and ravines. Before nightfall, Granger and Thompson had found the jaw of an unknown carnivore. Liu and Buckshot reopened a major fossil quarry first examined in 1925, an ancient streambed filled with skeletons of Eocene birds, fish, turtles, rodents, carnivores, chalicotheres (peculiar beasts with a head resembling a horse, teeth like a rhinoceros, sharp claws instead of hooves), and other mammals piled in a jumble like jackstraws. Separating the bones required enormous skill, and Andrews declared that "we obtained a more complete representation of the small fauna of this phase of the Eocene than had been discovered in any other place."

Until now, the weather had been ideal, with brisk nights tinged with frost and warm, clear days. Everyone knew such favorable conditions would not last, and true to their expectations, on April 24, the Gobi's wrath struck without warning. Heralded by twisting lines of wind devils, a furious storm raged throughout the night, filling the tents with dust and making sleep impossible. All the next day, the men struggled to extract more fossils from the quarry and plot geological formations while battling winds that swept the desert from every direction. Even stronger gales smashed the camp the second morning, keeping everyone pinned down under blankets, ripping or collapsing tents, and scattering clothing and equipment until late afternoon.

Taking advantage of a lull in the wind, Andrews, anxious to find a route westward into Chinese Turkestan, made preparations for a reconnaissance. He hastily departed with two cars on April 26, accompanied by Young, Hill, Shackelford, and Perez. Tserin went along to locate rendezvous points for the caravan. The local nomads were not optimistic regarding their chances, and from the outset, it was a hellish journey. Every trail seemed to lead nowhere. There were long stretches of jagged bedrock covered with yellow sand—absolutely impassable by automobile. And deadly traps of dry, crusted earth beneath which was a deep layer of jellylike mud into which Young's car suddenly "sank by the stern," Andrews recalled, "and came to rest with the engine pointing at the clouds." It required four hours to jack up the wheels on a foundation of stones and brush before the second car could pull Young's vehicle free.

Traveling for miles with only a compass and star sightings to guide them, constantly whipped by wind and sand, Andrews at last admitted "that we were enclosed by impossible country." After turning back, they were impeded by driving rain, thick fog, and fierce gales that lashed the explorers' faces until they bled, and sandblasted the cars' windshields so badly they had to be removed. They arrived at the base camp at Chimney Butte a week later, to discover that the rest of the party was faring no better. Everyone was huddled inside dust-shrouded tents with blankets and clothing over their faces and suffering from severely frayed nerves.

Despite the failure of the first attempt to reach Turkestan and the mysterious basins to the west, Andrews' staff was unanimous in the belief that they should try again. It was decided to use the same route taken by Sven Hedin's expedition, one that followed an old caravan trail and skirted the borders of Outer Mongolia, where foreign explorers were forbidden to venture. If Hedin had moved a caravan of 280 camels along this trail, it seemed reasonable to assume that it would be wide enough and sufficiently well worn for automobiles traveling in single file to navigate its surface. Although Hedin's objectives were not paleontological, Andrews "did not wish," as he had declared earlier, "to follow in his footsteps." Since there appeared to be no other alternative, Andrews sent Tserin and the caravan to the southwest toward Hedin's route—known as "the Great Mongolian Road"—that eventually led to Turkestan. Before Tserin departed, Andrews removed enough food and gasoline from the camels for three weeks and five hundred miles. The scientists planned to work a while longer among the rewarding fossil deposits near Chimney Butte, weather permitting, before joining Tserin at the first rendezvous.

But the Gobi's climate gave them only brief intervals of relief. Gales and sleet lashed furiously at the camp, and one sandstorm followed in the wake of another. Finally, Granger, Pond, Horvath, and Spock set out to explore promising fossil exposures to the east, but were forced to return the following day by unbearable winds. As far as their efforts at Chimney Butte were concerned, the explorers reluctantly admitted they were stymied. "The terrific winds made work in the fossil pits impossible," Andrews complained; "it had become very cold; the drifting sand and dust so filled our eyes that it was difficult to read or write; muffled in fur coats we could only sit in our tents and listen to the roar of the gale."

Andrews and Granger decided to abandon Chimney Butte and move to exposures in more sheltered recesses of the Shara Murun valley. On May 5, a beautiful, eerily tranquil day, the equipment was packed, the tents struck, and the hurriedly loaded cars were on their way by nine o'clock in the morning. As usual, Andrews and Shackelford were driving the lead car, considerably ahead of the main party. Just after crossing the Shara Murun River, scarcely more than a trickle due to lack of rain, Andrews wounded an antelope and walked over to kill it with the .38-caliber revolver slung over his left hip. In releasing a leather safety catch intended to prevent the weapon from falling out of its holster, his finger slipped and pressed the trigger. The pistol exploded against his left leg, inflicting a serious wound and knocking him to the ground. The bullet had entered midway along his thigh, projected downward, and emerged below the knee, nicking the distal end of the femur. Shackelford jumped in the car and raced to summon Dr. Perez and notify Granger to locate a suitable place to erect the tents. Appropriately, the spot Granger selected came to be known as "Hospital Camp."

After binding the wound and driving Andrews to camp, Perez operated on the leg that afternoon, assisted by Mac Young. Shu, one of the mess boys, stood by to help out as well, but the tears streaming down his face rendered him nearly useless. Using a heavy dose of morphine to kill the pain, Perez cut along the bullet's path, cleaned the wound, and skillfully stitched the torn muscle sheaths before applying a tightly bound dressing. As the morphine wore off, Andrews developed a high fever. In the worst of scenarios, a terrifying wind blasted sand and gravel under the edges of his tent, and for days, the air was a thick yellow curtain of dust. At night, the temperature dropped so low that ice formed on the water buckets. Even under these conditions, Perez managed to ward off any infection, and the injury began to heal so rapidly that within ten days Andrews could sit up in a chair whenever the wind died down.

"I think that no one who has not endured sandstorms," wrote Andrews, "can understand the torture to one's nerves, even in good health. Physically weak, in continual pain and with fever, it became well-nigh unendurable to me. Often I had to bury my head in the blankets to keep from screaming. It seemed that something in my brain would crack unless there could be a rest from the smash and roar of the wind, the slatting of tents and the smothering blasts of gravel."

CHAPTER 17

While Andrews was recuperating, scientific work continued whenever the blowing sand permitted. Although paleontological material was plentiful in the vicinity of Hospital Camp, extricating it was another matter. Often bones were no sooner laid bare before the wind brought the excavations to a halt. It was impossible to coat the fossils with gum arabic and paste them with flour-soaked burlap in a howling sandstorm.

On May 13, Granger took Hill, Shackelford, and Horvath eastward toward Iren Dabasu for a three-day reconnaissance. They returned greatly excited by the discovery that the sedimentary basin surrounding Hospital Camp extended to the east for at least another hundred miles and was rich with fossils. At the telegraph station at Iren Dabasu, they also learned that the Nationalists had captured Techow, only ninety miles from Tientsin and were pushing steadily north. No one knew whether all of northern China was about to erupt in warfare.

On May 19, Andrews paid a courtesy call on the priests at the nearby Baron Sog Lamasery to ask if he could store the fossils collected so far in one of the buildings until the expedition's return trip to Kalgan. In 1925, the lamas had rejected a similar request, claiming that evil spirits known as *"rinderpest"* had been angered by the removal of dragon bones from sacred ground. (*Rinderpest* was actually an infectious disease that attacked animals.) "After considerable talk with the head lama," Andrews mused, "and a promise of fifty dollars, he agreed the threat might be lessened if we kept it a profound secret among ourselves."

Mac Young had improvised a crutch made of sticks used to support Shackelford's portable darkroom. It allowed Andrews to hobble around fairly easily, and with the wounded leg healing rapidly, Perez consented to let the expedition start westward on May 20. Thirteen miles beyond a shrine known as Pailing Miao, or the Temple of the Larks—erected in veneration of the region's enormous flocks of larks that were highly prized by the Chinese for their exquisite songs—the caravan had stopped to await the cars. Tserin had hesitated to go any farther because of a Chinese *yamen* a few miles ahead that was apparently causing problems for other caravans. The camels had suffered terribly from lack of food since leaving the Shara Murun valley. Three had died and, after relieving the survivors of their gas and supplies, Andrews sent the weakest of them back to the lamasery at Baron Sog to be fattened for the return trip. The *yamen* proved to be a tax collection station between the provinces of Chahar and Suiyuan, where Andrews paid 91 cents for each of his camels and proceeded without incident.

Just at sunset that evening, a small caravan appeared over the hills. It was a detachment of Hedin's caravanners, most of whom had previously served with Andrews. After a joyous reunion, everyone sat in a circle in the twilight as Hedin's Mongols drew a crude map of their route in the sand. The journey had been a savage ordeal, they reported, and Hedin ran into serious political difficulties at the Turkestan frontier. Beyond the *yamen*, the drivers warned, the only food for camels was dried peas sold at a handsome profit by Chinese merchants who inhabited two mud huts along the trail. There was little water, and huge sand dunes, whipped by fierce winds, had blown across the trail. Hedin's caravanners were gaunt and hollow-eyed, having been forced to live on nothing but camel meat and small rations of dried peas. Several had nearly died from starvation.

Worse yet, the Communists had sealed off the "Great Mongolian Road" to Turkestan, which ran squarely through restricted territory inside the borders of the Mongolian People's Republic. All caravans were now limited to the Jao Lu, or Winding Road, a much more challenging route, virtually abandoned for centuries, that crossed some of the region's most inhospitable terrain. But Andrews was determined to continue westward. He believed that by going far enough south, he could circumvent the sand dunes and find fossil-bearing deposits.

Granger and the topographers—Hill, Spock, and their Chinese assistants—remained behind to map formations near the *yamen*, where they narrowly escaped being attacked one night by opium smugglers. Encircling their camp with four cars facing outward from the tents, they left the headlights on to illuminate the surroundings, and stood guard until daybreak. Andrews, meanwhile, had set out with the rest of the men and cars into a wasteland of black rocks, deep sand, and deserted temples. Freezing-cold nights were followed by afternoon temperatures of 145 degrees Fahrenheit, worse than anything Andrews had ever experienced in the northern Gobi. Nor was there any letup in the lashing winds and sandstorms that often brought the expedition to a halt. On May 29, an especially vicious wind arose out of the west and virtually destroyed the camp, sending equipment flying through the air, ripping tents, and forcing the explorers to remain buried under blankets and clothing for twenty-four hours. "There was something distinctly personal about the storm," Andrews related. "It acted like a calculating evil beast. After each raging attack it would draw off for a few moments' rest. The air, hanging motionless, allowed the suspended sand to sift gently down into our smarting eyes. Then, with a sudden spring, the storm devil was on us again, clawing, striking, ripping."

With the remaining camels becoming weaker and an uncertain journey ahead, Andrews had sent them back to a meandering stream called Haliu Ten Gol, about fifty miles west of the *yamen*, instructing Tserin to wait there.

As the cars fought their way toward the southwest, Andrews could scarcely believe the desolation that enveloped them:

The entire region was more arid and depressing than any that I have seen in Mongolia. It was totally unlike the Gobi as we had seen it north of the Altai Mountains. Each day the going became heavier and we were literally fighting for every mile. There were none of the wide spaces of the northern Gobi. To be shut in, oppressed by naked hard rock ridges without the majesty of size; to plow through drifting yellow sand in valleys neither wide nor narrow; to look upon the bleaching bones of camels, dead from thirst and hunger—that is worth enduring only when it is producing valuable results. Thus far we had obtained almost nothing and prospects for better success in the future were very dark.

In the midst of these trials, the wireless picked up a disturbing transmission which read in part: "... Chang Tso-lin retreated yesterday [into] Manchuria. The [Nationalists] are expected in Peking very soon. Everything normal in Peking and no trouble is expected. ...." Even so, Andrews was concerned, since all of the permits under which the expedition was operating had been issued by Chang Tso-lin. He feared that unless "adjustments," as he termed them, were negotiated by the American minister before his return to Kalgan, his papers might not be recognized by the new regime.

By now, it was also evident that the explorers were headed into a potential death trap. What appeared to be promising areas from a distance were impassable on closer inspection, and the dunes that lay across the caravan trails to the north were far more extensive than Andrews anticipated. Eventually, the cars had to be pushed for miles through sand, using heavy canvas runners reinforced with rope under the wheels. "To [proceed] with motors was absurd," Andrews entered in his journal; "only camels could carry on. . . . It was the end of the trail for us."

Returning dejectedly from their western fiasco, the explorers rejoined Tserin and the caravan encamped along the Haliu Ten Gol River. The camels were worse off than before. Five more had died, and the rest could hardly stand or walk, much less carry their usual four-hundred-pound loads. In a desperate effort to save them, Andrews purchased a large quantity of dried peas from merchants near the *yamen*, as there was practically no natural feed. He told Tserin to leave the camels at the river for at least another three weeks, and he would send cars to relieve them of their loads after the expedition established a base of operations in the Shara Murun valley. Andrews selected a location north of Hospital Camp on the edge of a high escarpment. It was near a spring-fed stream and overlooked a huge basin of Eocene and Oligocene formations.

While Granger and Thompson were engaged in excavating skulls of carnivores and titanotheres, a tattered Mongol hunter rode into camp armed with an antiquated flintlock rifle. Pond noticed that he was using a Dune Dweller scraper as the rifle's striker. The nomad claimed to have found it at a place twenty miles away where the ground was covered with chipped stones. Overcoming his initial fear of auto-

mobiles, he led Pond on a wild ride across the plains to a large site littered with artifacts. Scarcely had Pond begun exploring the area when
a foul-tempered lama drove him away, shrieking that it was "bad joss" (a
curse) to violate sacred places. Yet Pond's frustration was short-lived
when he stumbled upon an equally bounteous Dune Dweller station
north of Hospital Camp that yielded thousands of artifacts.

In the meantime, Horvath and Young returned to the caravan in
two cars to begin hauling gasoline and supplies to the camp. Aware that
the camels would be unable to travel for several weeks longer, Andrews
left on June 16 with Young, Spock, Hill, Perez, three Chinese, and two
Mongols to explore the country east of Iren Dabasu, a region that appeared as a blank space on existing maps. Approaching Iren Dabasu,
the pinion gear on one of the Dodges suddenly broke down. After
Young replaced the gear, they dropped off ten cases of gasoline at the
telegraph station to be picked up on their return.

Searching for an east-west trail, the explorers zigzagged back and
forth across the countryside until they came to a deep depression surrounded by bluffs. Evidently, it had once been an ancient lake bed, as
many of the exposed strata contained traces of fossils and thick layers of
fresh-water clamshells. The next morning, Buckshot discovered a partly
exposed mastodon skeleton and rhinoceros bones, which appeared to
identify the formation as Pliocene in age. Andrews was elated, since no
late Tertiary deposits—the period of fully developed mammals—had
been found in this section of the Gobi. Following a trail out of the depression, the men came to what Andrews described as "the largest and
flattest erosion plain . . . I have seen in Mongolia." But misfortune
struck again when the bevel or ring gear on the second Dodge gave
way. Since the only spare part was at the main camp, the car had to be
towed to the yurts of some nomads the scientists had visited earlier and
left there with two Chinese assistants until it could be repaired.

With seven men and extra gasoline piled into one car—and uncertain as to their exact location—they started back in the direction of the
base camp. Relieved to reach Iren Dabasu, Andrews found a telegram
from MacMurray with news of Chang Tso-lin's retreat toward Manchuria a fortnight earlier and informing him that the Peking government had collapsed. Tientsin and Peking had been peacefully occupied
by Shansi Nationalists. The telegraph operator also related that imme-

diately after Chang Tso-lin had abandoned Kalgan, a thousand bandits had swept into the city, looted numerous houses, and demanded a ransom of $500,000 from the Chamber of Commerce. They had fled two days later when an advance guard of Nationalist soldiers appeared.

At Andrews' suggestion, Granger had shifted operations fourteen miles north to Urtyn Obo, an area of deeply eroded exposures rich with fossils. At four o'clock the day after leaving Iren Dabasu, Andrews' party sighted the camp's tents floating in a shimmering mirage on the rim of a basin. Granger's blue eyes sparkled with excitement as he puffed vigorously on his pipe and led Andrews to the edge of a precipice overlooking a maze of pinnacles, spires, ravines, and chasms, streaked with shades of yellow, red, and gray. A sacred *obo*, or shrine, erected by the Mongols as an offering to the gods, stood atop a sentinel butte. There were so many fossils, Granger said, they could be seen from the tops of ridges with field glasses. Every bone stood out in stereoscopic relief, eliminating the need to explore the rugged terrain on foot.

"Success was immediate," Andrews raved upon first trying this innovative method of prospecting. Among a mass of bone fragments, his binoculars revealed the metatarsal of a *Baluchitherium*. Not far away lay the jaw and teeth of the carnivorous, piglike *Entelodon*. Urtyn Obo obviously had swarmed with a menagerie of other Tertiary beasts, especially rhinoceroses, the exotic titanothere *Embolotherium*, and enormous numbers of *Baluchitherium*. "So the story went," Andrews wrote excitedly. ". . . After the long weeks of discouragement, the days and nights of pain and heat and utter exhaustion, at last the desert had paid its debt. I slept that night with a great load lifted from my spirit."

Andrews pronounced Urtyn Obo—which the scientists named "Baluch Camp" because of its profusion of *Baluchitherium* remains— as "the most productive spot we had found in Mongolia," with the exception of the Flaming Cliffs at Shabarakh Usu. Every day, the paleontologists brought to light more evidence of the staggering number and variety of animals that had flourished there during the Tertiary period.

Andrews likened Baluch Camp to a "traveling zoo" because of the assortment of birds and mammals the men had acquired as pets and because of its bewildering array of extinct animals. Shackelford adopted two newly hatched horned owls, Perez had a falcon and a kite, Granger two golden eagles, Horvath a pair of ravens, Young a gazelle, and Buck-shot a hedgehog. Andrews was the possessor of a duck, a Mongol puppy, and his police dog, Wolf, who was always with him in the field. Andrews felt the presence of pets helped maintain the explorers' spirits. "Being out of touch with the rest of the world for months, as we were, anything that would give new interest and would take our minds off our work and each other was valuable."

On July 1, the temperature soared close to 140 degrees Fahrenheit in the sun and 110 degrees inside the tents, with the side flaps open and a breeze. The heat, absolutely unbearable by midday, lasted for three weeks without a break, except at night, when the thermometer dropped as much as 70 degrees. When the caravan finally lumbered into Baluch Camp, it was apparent that the heat had devastated the already weakened camels. Six more had died and several others were in terrible condition. After reloading the boxes, Andrews dispatched the weakest camels to Eriksson's mission at Hatt-in-Sumu, while Tserin took thirty-seven of the stronger camels, picked up those previously sent back to the lamasery at Baron Sog, and headed for a temple called Bulga-in-Sumu, east of Iren Dabasu, to await the cars.

Meanwhile, the unrelenting heat was causing a serious loss of gasoline. Even though the storage tins had been reinforced with extra solder along the seams and packed in heavy wooden cases, the 70-degree difference between day and night temperatures caused many of the tins to burst from expansion and contraction. "During the night there were continual explosions like fire-crackers," wrote Andrews, "which meant other empty tins in the morning." A thousand gallons had been wasted, equaling the loss of hundreds of miles of travel.

From Baluch Camp, the expedition moved to a location thirty-six miles away that was dubbed "Holy Mesa." Sitting atop an escarpment was a large *obo*; at its base, nomads had erected a smaller *obo* partially made from *Baluchitherium* bones. Five days of prospecting in the area confirmed the existence of rich fossil deposits, but Andrews, anxious to push east, decided to postpone further excavations at Holy Mesa until another season. In any event, following the hottest night of the summer and a ferocious sandstorm, a procession of lamas appeared to protest that the gods were offended by the strangers' presence; the mesa, they said, was too sacred even for nomads to graze camels there.

Startling news awaited Andrews when his procession again reached Iren Dabasu. MacMurray had sent word that China's capital had been moved to Nanking, Peking was now renamed Peiping, and control of the northern provinces was being uneasily shared by Chiang Kai-shek, Feng Yu-hsiang, and Yen Hsi-shan. Andrews, however, was more concerned at this point about his dwindling supply of gasoline and the insufferable heat that threatened the continuation of explorations.

Stopping at their old camp near Iren Dabasu, Granger took some of the scientists to a spot where he had found fragments of eggshells in 1923. Scarcely two feet under the surface, Granger exposed several nests, each containing fifteen to twenty-five eggs in circles of two or three layers. Undoubtedly dinosaurian, they were slightly round in shape and almost white—quite different from the eggs at the Flaming Cliffs. Granger surmised that these were laid by duck-billed iguanodonts, which had flourished here in enormous numbers during the Cretaceous period. While waiting for Tserin and the caravan to arrive, the cars moved a short distance eastward to an escarpment overlooking a broad basin. A stroke of luck caused the men to camp in the midst of an area littered with fossils. Immediately, Buckshot was hard at work uncovering the skeleton of a small mastodon. It appeared to have been pulled apart by predators, but the bones were well preserved.

Judging by the thick deposits of freshwater clamshells, "Mastodon Camp," as this new location was named, was situated on the edge of a large ancient lake. The surrounding exposures were among the most recent yet encountered. Spock dated them as Pliocene to Pleistocene in age—approximately 5.2 million to 1.64 million years old. "Everything we found there," noted Andrews, "would . . . throw a brilliant light upon what had been the [least-known] period of Mongolian life. . . . [Also] it was just the geological age in which we might look for early human types with greatest hope of success."

While at Mastodon Camp, Hill appeared one day with a baffling discovery: a flat, platelike object about ten inches long, nine inches in width, and three-quarters of an inch thick. Its enamel covering left no doubt that it was a tooth of some unknown creature; three others were later found, but they offered no clues as to what kind of animal it represented. For the time being, it would remain a puzzling mystery.

At last, the debilitating heat was broken by heavy downpours that blew in every afternoon around five o'clock from the northwest. The rains drenched the desert, sending water pouring down the hillsides, surging through ravines, turning depressions into shallow lakes, polluting wells with silt, and even sweeping away a nomad settlement, leaving ten of its inhabitants drowned or hopelessly buried under mud.

Because some of the fossils were too large to be safely loaded onto the camels, Andrews, Young, Hill, and Spock took two cars piled with specimens to Hatt-in-Sumu on July 22 for storage. Everyone was relieved to learn that General Yen Hsi-shan's soldiers had almost completely rid the plateau of brigands, allowing normal trade to resume. Andrews reported seeing a stream of caravans and perhaps two thousand carts on the road out of Kalgan to Ulan Bator and other western trading centers. At P'ang Kiang, they were even able to buy eight cases of gasoline.

Struggling through fifty-five miles of thick mud after leaving P'ang Kiang, the cars arrived at Hatt-in-Sumu, where they unloaded the massive fossils. Departing the next morning, the explorers bought three more cases of gas at P'ang Kiang, then turned north and east toward Mastodon Camp. During their absence, Tserin and the caravan had arrived, but Andrews was horrified to learn that twelve of the twenty-four remaining cases of gasoline had burst. Transferring the rest of the gas to the cars, Andrews sent the camels, bearing almost ninety boxes of fossils and other specimens, to await the cars at Hatt-in-Sumu.

From Mastodon Camp, the expedition moved to a bluff above a basin known as Tairum Nor. Ancient graves, sealed with heavy rocks, were scattered at the base of the escarpment, and Pond, assisted by Shackelford, Hill, and several Mongols, opened one of the largest of these tombs. Inside was a well-preserved skeleton, lying facedown. Originally, the body had been clothed in a robe or dress, probably woven of fiber, and adorned with beads made of freshwater clamshells. But in terms of fossils, the camp at Tairum Nor was a disappointment, and after five days of prospecting, Andrews gave orders to leave at daylight the following morning. With his usual thoroughness, Granger decided to examine one last unexplored exposure. On his way back, he and his two Chinese assistants stumbled upon bones of what first appeared to be a mastodon. After examining the jaw, they realized that the mystery of the strange flat, teethlike plates found earlier by Hill had been solved. Andrews described the spatulate front of the lower jaw as "resembling a great coal shovel. Side by side, in the distal end were two flat teeth like the others we had found. They were eighteen inches across. Behind them, the jaw narrowed and then divided into the two branches which bear the molar teeth. The jaw was five feet long—a perfectly stupendous organ."

It was clear that the two shovel-like lower teeth had been adapted for scooping up food, most likely succulent water plants from along the shores of streams and lakes. Using an elongated upper lip, the animal pushed the food into the back of its mouth to be ground up by the molar teeth. Later examination at the Museum established that the creature was a larger species of a related genus known as *Platybelodon*, and it was almost identical to specimens from Russian Turkestan and Nebraska. Although the body and legs resembled a mastodon, the head, with only rudimentary tusks and no trunk, was strikingly different from other mastodons and elephants.

Despite the rapidly dwindling supply of gasoline, the expedition continued eastward in search of promising areas for next season's explorations. Such enticements were everywhere—eroded landscapes covered with bones of dinosaurs and mammals, extensive Dune Dweller sites marked by thousands of artifacts, seemingly limitless expanses of unmapped desert, and unexamined geological formations. Even from the most superficial observations, Andrews was certain these eastern portions of Inner Mongolia held undreamed-of scientific treasures.

Nomads directed the men to a small lake, Koko Nor, next to a collapsing temple occupied by a dozen impoverished Chinese Muslims from Kansu. Over half were suffering from advanced stages of syphilis, and although Perez did what he could to treat them, there was little hope they would survive. They told of a much larger lake, Hul Tsagan Nor, farther to the east, but the route was blocked by mud and sand dunes. As the countryside was a mosaic of Pliocene sediments, perfect for fossils and traces of early hominids, Andrews resolved to explore this lake district by camelback in 1929.

By now, the strain of driving through miles of sand and rock had nearly exhausted the gasoline supply, forcing the cars back to P'ang Kiang. Andrews telegraphed MacMurray at the American legation in

Platybelodon

An elephant-like mammal with two broad lower teeth resembling a shovel. These enabled the Platybelodon *to scoop up vegetation along riverbanks, lakes, and marshes.*

Peking that he and his companions would be entering Kalgan with passports issued by Chang Tso-lin, who had been killed en route to Manchuria when his train was bombed. Surprisingly, MacMurray had no difficulty arranging the validation of Andrews' documents by Yen Hsishan's troops, thus allowing the expedition free access to all border and customs *yamens*. After meeting Tserin at Hatt-in-Sumu, Andrews unloaded and stored surplus equipment for the next season, and paid off most of the Mongol drivers and assistants. He then sent Tserin to Kalgan with fifty camels, weighed down with eighty-seven boxes of specimens. The eight Dodges, carrying the oversized crates, left Hatt-in-Sumu on August 11 in a raging blizzard and subfreezing temperatures.

Bypassing Changpeh, the cars drove directly through the Wanchuan Pass to Kalgan late that afternoon. When they finally negotiated the deeply rutted road to the bottom of the pass, it was dark and a hard rain was falling, making the last few miles of road into Kalgan dangerously slick. At one of the worst spots, the car driven by Perez became badly stuck. Horvath decided to attach ropes through the spokes and around the wheels for added traction. Unable to see well in the darkness, Horvath's knife slipped and plunged into his thigh, severing an artery. By the time Perez could get a tourniquet around the leg, Horvath was soaked with blood and extremely weak.

Arriving at a military post on the outskirts of Kalgan, one of the guards raised a minor objection to the men's passports on a technicality. As Alonzo Pond later recalled in his book, *Andrews: Gobi Explorer*, Andrews was patient until the guard ordered them to remain where they were until morning.

"'You,' said Roy in the quiet voice of authority, 'will step aside and let my cars and my personnel into Kalgan according to our authorization and the order you have received from Peking, or we will shoot our way through.'

"The threat of force presented in official Mandarin Chinese with the self-confidence Roy Andrews could command outfaced the guard. He finally realized that holding the [explorers] would cause him to lose face with his superiors and might bring discipline instead of praise." The cars drove to the British-American Tobacco Company, where they were given quarters and Horvath's leg could be properly treated.

Getting to Peking proved to be a nightmare. Before his departure, Chang Tso-lin had confiscated virtually every railroad car. The only train between Kalgan and Peking sometimes ran once a day or once a week, and was always crammed. There was no choice but to drive the 145 miles over a road designed for mules and wagons. Only two or three cars had attempted the trip, and they were empty. Andrews' cars were grossly overloaded. For three days, the men fought rain, mud, rock slides, and torrents of water before reaching Peking. One car broke an axle. Another sank into deep mud, and others were filled with water and almost swept away. "How anything on wheels," Andrews marveled, "designed to travel on a road, could stand the punishment those cars received was beyond my conception. But they kept steadily on . . . [and] we passed through the great gate of the Tartar Wall at eight o'clock on the third night after leaving Kalgan."

Regardless of the expedition's failure to penetrate the western desert leading to Turkestan, the crippling shortage of gas, and the constant onslaught of sandstorms, rain, and extremes of heat and cold, Andrews was "well content" with its results. The paleontologists had added significantly to their knowledge of the Gobi's fossils; ten thousand

Dune Dweller artifacts were collected; a half-dozen new geological formations were identified; and some three thousand miles of unknown territory had been partially mapped. Equally important, Andrews had turned up promising locations for future explorations in areas where now-extinct animals once flourished.

CHAPTER 18

Ten days after returning to Peking, Andrews received word that Tserin would shortly arrive in Kalgan with the eighty-seven boxes of fossils. Articles immediately began appearing in foreign and Chinese papers reporting that the Society for the Preservation of Cultural Objects had ordered the governor of the Chahar district to detain the expedition's specimens at Kalgan. SEIZURE OF FOSSILS FACED BY ANDREWS, screamed a headline in the *New York Times*. Released by the Associated Press, the story was picked up by newspapers around the world.

Granger and Young caught the first train to Kalgan, only to learn that the caravan had been met at the top of Wanchuan Pass by soldiers, who escorted them into Kalgan. To avoid the danger of looting, Granger prevailed upon the commissioner of foreign affairs to transfer the crates to Larsen's compound under armed guards. In the meantime, the Cultural Society and another of their official branches, the Commission for the Preservation of Ancient Objects, bombarded Chinese papers with the usual barrage of antiforeign accusations. They charged that the Central Asiatic Expeditions had "trespassed upon China's sovereign rights" and "stolen priceless treasures"; that the scientists were in reality "spies against the government"; and they had been "searching for oil and minerals and smuggling opium."

Andrews was staunchly defended by the prestigious Geological Survey of China, especially by its respected paleontologist Amadeus W. Grabau, who protested the Cultural Society's position in a formal document he sent to its members. In addition, editorials favorable to the Asiatic Expeditions appeared in the *North China Standard* and the *Peking Leader*, which published a defense of the explorers' achievements and drew a sharp distinction between the removal of fossils and zoological specimens from Chinese soil, and the theft of art treasures and material of historical importance.

Accompanied by the secretary of the American legation, Andrews called on General Yen Hsi-shan's representative to review the issue. But their efforts were to no avail. A few days later, Yen notified Andrews that it was a matter for the minister of foreign affairs in Nanking and he refused to intercede.

On September 11, an extremely frustrated Andrews wrote to Osborn:

> I am in the middle of a battle royal with the Chinese. If they win out and keep all our things we might as well quit now and close up here. If we win we can go on. I am *outraged* by the charges they have made against us and for my own dignity and that of the Museum, cannot deal with the Society. It has no official standing and the Legation will have nothing to do with it. . . . They want "squeeze" of course. They held Sven Hedin for $30,000. I'll not give them a cent—anyway it would cost us at least $50,000 and would make us lose such "face" if I had any dealings with them now, that we'd have to quit.

Andrews was determined to pursue his cause to the highest diplomatic levels if necessary. In two letters, he suggested that Osborn contact the secretary of state in Washington, Frank B. Kellogg, and persuade him to use the full weight of his office to break the deadlock. Nor did Andrews rule out taking the conflict directly to the president, Calvin Coolidge, as a last resort. "Of course," Andrews told Osborn, "I can't leave [China] until the matter is settled. . . . I am so disgusted and angry that perhaps I ought not write to you now but I must 'let off steam.' I'll do my best but I am afraid that we are in for a real row which may last for weeks."

Not long thereafter, Andrews began to realize the full impact of the Cultural Society's obstructive tactics, and the diplomatic snarls that would result from the confiscation of what the world's press widely touted as "Andrew's Eighty-seven Crates." Publicly, the Museum's representatives and United States government officials were obliged to handle the issue with utmost delicacy. It was soon evident that they were walking a thin line between Chiang Kai-shek's Nationalist regime in Nanking, which appeared to hold the key to resolving the problem and was represented by China's ambassador in Washington, and the Cultural Society's members—bolstered by Communists, intellectuals, and student agitators—who viewed the Nanking clique with contempt and exerted a powerful influence in northern China.

Yet privately Andrews, Granger, and Osborn fumed with indignation at both factions. No one could be trusted; bribery and deception were rampant; and the Cultural Society had divided into a bewildering number of splinter groups. One was forced to deal with such entities as the Commission for the Preservation of Antiques, the Chinese Historical Preservation Committee, the Association for the Preservation of Ancient Objects, the Society for the Maintenance and Preservation of Cultural Objects, and the Commission for the Preservation of Ancient Objects. Often, name changes were hastily made to conform with a particular situation. For example, when Andrews argued before the Society for the Preservation of Cultural Objects that fossils—buried in the earth millions of years before humans appeared—could not possibly be considered "cultural" objects, the organization promptly altered its name to the Society for the Preservation of Ancient Objects!

On September 12 and 13, the governor of the Chahar district and other officials, acting on orders from the Cultural Society, opened and inspected the contents of Andrews' eighty-seven cases at Larsen's compound in Kalgan. Just as Andrews had avowed all along, they contained nothing but "bones and horns of birds and animals," plus some old clothing. No opium was found, thus refuting the society's senseless allegations of drug smuggling.

In the meantime, the society issued an edict that anything of archaeological importance—including fossils—was to be confiscated. (Unknown to the Chinese, Alonzo Pond had already shipped the Dune Dweller artifacts in steamer trunks from Tientsin as part of his personal baggage.) Duplicate paleontological specimens, the society ruled, could be released to Andrews as "gifts," but all unique fossils must remain in China. Suddenly, however, in a complete volte-face, the society arbitrarily declared that no fossils of any kind could be separated from "ancient cultural products" and therefore were forbidden to leave the country even as gifts.

After a month of haggling in Kalgan and repeated changes in regulations imposed by the society, the collection was finally transported to the expedition's headquarters in Peking, where the crates were subjected to a more detailed inspection. Believing there would be no further objections to the removal of the collection, Andrews instructed Granger and his Chinese assistants to begin conserving the fossils in the headquarter's laboratory and reducing the number of crates from eighty-seven to thirty-seven for shipment to New York.

Andrews departed for the United States on the Trans-Siberian Railway, stopping in England to visit Yvette, George, and Kevin. He arrived in New York on December 10, where he conferred with Osborn regarding future plans and spent a month lecturing on the East Coast and playing polo. He also found time to write articles for various magazines, including a series for the *Saturday Evening Post*. In addition, he had completed an autobiographical memoir for George Putnam, *Ends of the Earth*, which was published in 1929 and reprinted two years later in the *Illustrated London News*. On January 21, 1929, at a gala ceremony, Andrews received the Elisha Kent Kane Medal for outstanding achievements in exploration from the Geographical Society of Philadelphia. But writing to Granger earlier in January, he had confessed, "I am not enjoying New York. . . . The wheels of the city seem to revolve about fifty times faster than they did when I was here last. I am sick to get back to China!"

Following a whirlwind lecture and polo tour of the West Coast in February, Andrews sailed for China aboard the *President Cleveland* to finalize his strategy for the 1929 expedition to Inner Mongolia. He arrived in Peking on March 23 to find conditions far more precarious than he had expected. The Cultural Society was once again in a fury over plans for the expedition, and had drawn up a list of conditions governing its activities. Most objectionable were the society's demands that half of the expedition's members had to be Chinese; that at least one Chinese codirector be appointed; that all fossils must remain in China, except for rare specimens that required expert conservation at the American Museum; all such fossils were to be accompanied to New York by Chinese scientists whose expenses would be provided by the Museum; and after the restorations were completed, these items would be returned to China.

What transpired next was an endless series of negotiations involving Andrews, Osborn, Granger, the Cultural Society, the American Museum, the State Department, the Chinese ambassador in Washington, Chao Chu Wu, the Nanking government, and the American legation in Peking.

Early in the spring, after weeks of sitting on the docks at Tientsin, twelve of the cases prepared by Granger had been released and shipped to New York by way of the Suez Canal. Inexplicably, no one at the Museum, in Nanking, or Washington was notified of their departure until months later, when they reached New York. Soon thereafter, the other twenty-five boxes of fossils arrived in Tientsin, but the Cultural Society refused to allow their export.

Meanwhile, Osborn had drawn up his own demands for proceeding with the 1929 expedition, countering the terms previously outlined by the Cultural Society. His conditions included the immediate release of all specimens being held by the Chinese (he was still unaware, of course, that twelve crates were en route to America), and the removal of any obstacles preventing the scientists from working in Inner Mongolia, with a further promise of no interference from the Cultural Society concerning the shipment of fossils out of the country. Under pressure from the State Department, Ambassador Wu reluctantly agreed to contact Nanking with Osborn's proposal. But Nanking's only offer of help was to suggest introducing a bill in the Nationalist Legislature, the Yuan, during the summer—too late for the expedition—that would clarify a major point of contention: the hotly debated distinction between what Osborn called "artistic works of man" and "natural works of nature." It was suggested that such a ruling might resolve the issue of whether China or foreign scientists could rightfully claim ownership of fossils.

The battle over the crates being held in Tientsin raged until midsummer—despite the ludicrous fact that the other twenty-five had *already* been shipped on June 2. Andrews had managed to bribe an official of the Cultural Society named T. T. Sun, who secretly ordered their release. But Andrews' coded message to Osborn that "they are on the high seas, thank God!" failed to reach the Museum for eight weeks, leaving all the parties involved totally in the dark and embroiled in a comedy of errors. Oblivious to the clandestine arrangement between Mr. Sun and Andrews, the Cultural Society, the Museum, the Nanking government, Ambassador Wu, and the State Department went right on wrangling over the fate of *all* thirty-seven crates.

Osborn and the Museum's new director, George Sherwood (who had replaced Lucas in 1924), enlisted the aid of the newly appointed secretary of state, Henry L. Stimson, and Assistant Secretary Johnson in their ongoing struggle with Chinese authorities regarding the terms of the 1929 expedition. To Granger's dismay, Andrews agreed to take along two Chinese assistants at a salary of $150 a month. "If they must save their faces by saying they are going to be co-Directors," he told Granger, "let them say so, only let them have it thoroughly understood that you will do the actual [scientific] directing." Granger was not at all happy with this arrangement, and in a rare outburst of anger, he referred to the project in a letter to a colleague at the Museum as "Roy's Charity Organization." Furthermore, Osborn had agreed by then to the Cultural Society's demand that a Chinese student be brought to New York and trained in paleontological research at the Museum's expense. But the society's insistence that all fossils from any future explorations must be kept in China, except under the conditions they had previously outlined, was unacceptable.

Disgusted by the endless obstacles thrown up by the Cultural Society, Andrews threatened to pack everything, leave China permanently, and terminate all scientific cooperation. In a cable to Osborn, he ranted:

> Chinese attitude extremely unfriendly. Will have good effect if Secretary of State for Foreign Affairs informs Chinese Ministers that stopping work will result in . . . unfavorable American public opinion.

Osborn was desperate that Andrews be allowed to "clean up" the sites previously found in eastern Inner Mongolia. He instructed Andrews to keep the expedition ready to enter the desert, but Andrews felt that if he waited longer than June 15, there would not be adequate time to carry out worthwhile explorations. At one point, Osborn and Secretary of State Stimson considered asking President Herbert Hoover to intervene. And in a vigorous display of resolve, Osborn quoted Napoleon in a letter to Andrews: "Never defend yourself; always attack," to which he added, parenthetically, "When you are conscious of being entirely in the right."

In late June there was an apparent breakthrough when Ambassador Wu agreed to put the Museum's conditions in writing with his full compliance. Osborn was informed by Wu on June 20 that the Nationalist government in Nanking had replied by cable, saying that even though the Cultural Society would *not* release the crates—which unbeknownst to anyone except Sun, Andrews, and Granger, were already en route to New York—Nanking would agree to allow their shipment! Wu admitted to Osborn that he felt "somewhat embarrassed" by his government and reversed a statement he had made to Osborn characterizing the Cultural Society as "friendly but misinformed" to "unfriendly and misinformed."

Equally unexpectedly, the minister of foreign affairs in Nanking, a man named Wang, suddenly agreed (or appeared to agree) to some of Osborn's demands. Referring to the Cultural Society as "inexperienced," he arranged a meeting between Andrews and the Society in Peking, explaining that even though the Society had not changed its position, it would consent, in a curious ploy to save face, "not to bring up the disagreements." Unsure of the reason for the meeting, Andrews and the American minister, J. V. A. MacMurray, appeared, only to find that the two-faced Wang had been there ahead of them to assure the Society of his vehement support for their hard-line policies, including the refusal to release the fossil crates—an act of betrayal that prompted Andrews to denounce Wang as "a traitor" and the "worst of the lot."

Seething over the Cultural Society's unending hostility, Osborn complained to Ambassador Wu, "Why does China continue to offend, delay, disappoint, and obstruct . . . when she needs U.S. cooperation? Why is China not friendly like the rest of the world?" To Andrews, he wrote, "This had been a terribly long and trying experience . . . let us not surrender our territory nor our rights. It is better to go [into the desert] a short time this season than not go at all."

But Osborn's hubris proved fruitless. Andrews canceled the expedition on June 26. He sent an exasperated letter to Osborn stating that the Cultural Society had backed itself into a corner so tightly that the society would obviously lose face if the expedition were to proceed. "None of you in America," he declared, "understands how little the Nanking government has to say about things in the north. . . . [The] Cultural Society, in our last interview, gave us to understand that unless we submitted to their demands they would prevent us from going. They need only tell the police to keep us from shipping our cars."

Andrews informed Osborn that he was sending the ten American and European staff members back to their respective homes. He planned to remain at the headquarters in Peking to begin working on what would turn out to be his magnum opus, the triumphantly titled—some have called it chauvinistic—*The New Conquest of Central Asia.* Ever the optimist, Andrews had not ruled out returning to Inner Mongolia in 1930 if political obstacles could be overcome. He also suggested to Osborn the possibility of joining Russian scientists in carrying out explorations in Russian Turkestan. Osborn seemed eager to pursue this idea as an option and immediately contacted his friend William Morden, who had led zoological expeditions for the Museum to Siberia and central Asia, including Turkestan, and was well connected with Soviet scientists.

Yet Osborn's obvious preference was to see the resumption of Andrews' Mongolian explorations. As he had emphasized to Wu during their protracted negotiations, the Museum was obligated to finish the project to which its world-famous financial benefactors and the American public had thus far contributed over $500,000. Osborn was equally anxious to witness the completion of what he called Andrews' "great life work." "Let us never acknowledge that we are beaten," he wrote to Andrews on July 17, refusing to concede that China should ever be allowed to *"arrest the work of the most important natural history expedition of the Twentieth Century."*

At the end of July, Osborn finally received Andrews' messages that the crates were safely en route. Osborn notified Andrews on August 19 that he was taking the following steps to organize another expedition in 1930: sending a letter through the assistant secretary of state containing six points for Secretary Stimson to discuss with the Chinese; arranging an appointment with President Herbert Hoover "in order to place the matter before his mind"; consulting with J. P. Morgan, Jr., and Thomas Lamont to make further loans to China "conditional on continuation of the Central Asiatic Expedition"; setting up a meeting with Hoover,

Stimson, Johnson, Assistant Secretary of the Treasury Mills, and George Otis Smith of the United States Geological Survey. "If they decided to continue the Central Asiatic Expedition," Osborn pronounced with imperialistic fervor, *"it will certainly go through."*

By September 20, Andrews had regained his equilibrium after the stress of the previous six months, and hastened to put Osborn at ease in a revealing letter.

> When I wrote you before, I was deeper in the slough of despondency than I have ever been in my life. That was quite naturally due to the physical evidence of the break up of the expedition—the departure of Young, Granger & Thompson, the sending away of our Chinese staff who have been with us for so long and all the other evidence that the end had come.
>
> Now I have come up out of that and am really happy again. I am intensely interested in Vol I [*The New Conquest of Central Asia*] which is progressing rapidly. Thirty six chapters have been completed! Polo gives me change from work, recreation & exercise. I have four of the best polo ponies in Peking, fast, handy & quick to respond to training. My life has settled into a very satisfactory routine. I am physically well & and mentally fairly happy because every day I can see progress in Vol I and realize so fully that this is the only place I could do it. Therefore, dear Professor, put your mind at ease concerning me. I am utterly ashamed of myself for having unloaded my unhappiness on your shoulders. . . . But it is all over now & my spirit has climbed back to its accustomed place on the rosy cloud where it can look down into the black deeps and laugh because I once sunk so low. . . . Thanks for your affectionate understanding. I've decided not to die yet.

CHAPTER 19

While Andrews watched the unfolding events in China, he and Osborn intensified their efforts to organize a combined American Museum–Russian expedition to Turkestan. Aided by William Morden and Osborn's friend A. Borissiak, contact was initiated with the Academy of Sciences in Leningrad regarding logistics and specific objectives. Simultaneously, tantalizing reports began arriving from Joel Eriksson describing promising fossil deposits observed by a nomad informant along the shores of a lake in the sandy wastes of eastern Inner Mongolia, the area Andrews had wanted to explore by camelback in 1929. Excited by Eriksson's news, Andrews and Osborn decided to make an all-out effort to regain access to this region before abandoning further work in China.

But the Cultural Society remained as truculent as ever. Once more, they had blocked Sven Hedin's plans to launch another expedition—this time to Kansu Province—for five months until a succession of obstacles could be cleared. Sir Aurel Stein, one of the most renowned and controversial of Asian archaeologists (who had justifiably incurred the wrath of the Chinese by looting ancient manuscripts and works of art), would shortly have his passport revoked by irate authorities egged on by the Cultural Society. Even Davidson Black, the respected anatomist at the Peking Union Medical College, was fearful that the society might confiscate the bones of Peking Man, which Black had been studying for four years. In short, the climate of hostility toward foreign explorers and scholars was rapidly worsening.

In July of 1929, Andrews had written to Osborn that he saw little hope of resuming explorations. "There is just one way in which something might be done," he remarked. "That is by unfavorable publicity.

The Chinese are very sensitive to criticism and ridicule. . . . [And] I feel that statements from you to the press are necessary." Andrews' belief in the power of publicity stemmed largely from an article he and Osborn had published in the September 27, 1929, issue of *Science*, entitled "Interruption of Central Asiatic Explorations by the American Museum of Natural History." Osborn had proclaimed, "It is one of the most regrettable incidents of recent times. . . . The purely scientific and educational aims of other countries are misrepresented in the Chinese press, and popular opinion is inflamed against scientific exploration as hostile to the best interests of China." Andrews, in turn, characterized the cancellation of his explorations as "a calamity in the advance of science" — a sentiment widely supported in newspapers throughout the United States and Europe, to the aggravation of the Cultural Society.

By the autumn of 1929, an unexpected turn of events offered Andrews a glimmer of hope. General Yen Hsi-shan and the Christian zealot General Feng Yu-hsiang began a concerted effort to topple the Nationalist regime in Nanking, eliminate Chiang Kai-shek, and restore Peking as the capital. Andrews was certain he could deal with Yen Hsi-shan in curbing the Cultural Society's power, though he was less sanguine about working with his longtime adversary Feng. Yen Hsi-shan had been cooperative in the past, and he desperately wanted to generate a more positive image of China in the foreign press.

Then Andrews' hopes evaporated. In typical warlord fashion, Chiang Kai-shek raised millions of dollars, bought off the leading generals of his opponents' armies, and maintained the status quo in Nanking, leaving the Cultural Society's power completely untouched. An even worse blow had fallen several months earlier when the Nationalist legislature, the Yuan, finally passed a bill governing the export of antiquities. Its provisions were far more stringent than anyone anticipated: no ancient or historical material — including fossils — would be allowed to leave the country under *any* circumstances. And to Andrews' dismay, the new restrictions were passed with the full backing of various organizations that had formerly supported his endeavors: the Geological Survey, the National Research Institute, the Peking Society of Natural History, the Library Society, and nearly every other scientific and educational entity in the country.

Stunned by Nanking's actions, Andrews again solicited the help of his former ally T. T. Sun in a risky plot. It was Sun whom Andrews had bribed to secure the release of the last twenty-five crates of fossils from the 1928 expedition. Although he was a member of the Cultural Society, Sun was a diffident, money-hungry man, utterly lacking in nationalistic fervor and easily manipulated. Sun advised Andrews that his only hope of returning to Inner Mongolia lay in the art of subterfuge. He advised Andrews to sign an agreement abiding by the Yuan's newly enacted laws. Then, with Sun's complicity, he should negotiate an understanding—"under the table" and involving the payment of substantial squeeze—that would allow Andrews to remove fossils in accordance with the terms of his earlier agreement: duplicate specimens would be given to China, along with casts of unique items. He also offered outright to act as an agent on Andrews' behalf, reporting on everything that transpired during the Cultural Society's meetings— providing, of course, he was paid enough to betray his own organization and champion the expedition's cause.

In a coded message to Osborn, Andrews related information obtained from Sun to the effect that "the Cultural Society are sorry that the expedition has been stopped & he believes they would be open to new negotiations. . . . Mr. Ma [a high-ranking official of the society with dubious scruples where squeeze was involved] . . . said that the severe conditions they imposed were to establish a precedent and 'face' for themselves; that the conditions as to the determination of the duplicates, which we would not agree to [previously], would not be insisted upon *in actual fact.* [Mr. Sun] strongly urged that we sign the 'conditions' [passed by the Yuan] to give the Cultural Society 'face,' with the verbal understanding that we would be allowed to ship our specimens to New York and retain originals as in the past. . . . I believe that the Cultural Society has been considerably chastened by the unfavorable publicity which they have received. . . . The *Science* article has had much effect & they are beginning to realize that they made fools of themselves. They want to 'save their face' and still let us go out."

All through the winter of 1929–1930, Sun played his role as a "spy," keeping Andrews informed of the Cultural Society's deliberations by notes delivered to the expedition headquarters and picking up his hush money in envelopes that he always burned. To further protect himself in case his payments from Andrews should be discovered, Sun's financial requests were invariably couched as dire personal calamities. One letter told of ruinous debts incurred when he spent nearly twenty days in a hospital recovering from a terrible sickness. Another pleaded that his salary from the commission was causing his financial condition to go "from bad to worse." But most frantic of all was a famine that allegedly struck his district, leaving nothing for his family and neighbors to eat except roots and leaves. "[So] I hope you . . . make a good contribution to save them."

Eventually, the well-paid Sun's maneuvering within the Cultural Society's inner circle—and his judicious distribution of squeeze—was successful. Several months of wrangling ensued before an agreement was drafted, during which Andrews' dealings with the society could hardly be termed cordial. Nevertheless, by March 18, 1930, he had once again accomplished the seemingly impossible. He had obtained permission to launch another expedition to Inner Mongolia.

Under the terms of Andrews' permit, he was not allowed to collect zoological or archaeological material. Explorations were limited to paleontology, geology, and topography, which restricted the composition of the scientific staff. Granger, Thompson, and Young arrived in Peking at the end of April. A. Z. Garber, of the Peking Union Medical College, joined the expedition as surgeon; and Andrews' friend Lieutenant-Colonel N. E. Margetts, an attaché at the American legation, arranged for the services of Lieutenant W. G. Wyman as the topographer. (Like Butler in 1925, Wyman prepared an intelligence report on his observations in Inner Mongolia for the War Office in Washington.) Doctors C. C. Young and S. C. Chang were delegated to represent the branch of the Cultural Society known as the Commission for the Preservation of Ancient Objects.

Anxious to secure film of the planned foray by camelback into the eastern sand dunes, Granger wrote to Shackelford in Hollywood asking if he would rejoin the expedition. Back came a reply bursting with Shackelford's irrepressible enthusiasm: "Would I be willing to go?!!!! does a duck swim? does the sun ever shine in the Gobi? does argul make a good fire? Why Walter Granger! would I be willing?? Now if you had asked 'Can I go' what a different story . . .

"Last year . . . was the worst I have gone through, financially, for a long time. I am so deeply in debt that I don't see how I can get away ahead of the sheriff." Andrews, with his expedition strapped by unexpected costs and unable to raise additional funds because of the Depression, found it impossible to meet Shackelford's needs. So the venture was forced to proceed without a photographer.

In selecting a geologist, Andrews was more fortunate. He was able to enlist the world-renowned Jesuit scientist Père Teilhard de Chardin, who had spent years in China tracking human fossils and studying geology. Teilhard's inclination to interpret theological questions such as "original sin" in the context of evolutionary theory had resulted in his expulsion as a lecturer at the Catholic Institute in Paris. In 1926, he went into virtual exile in China, where he served as adviser to the Geological Survey. In 1929, he became chief geologist for the Peking Man excavations, and gained considerable recognition for his work with Davidson Black in analyzing the skeletal material from the site. Osborn, Granger, and Teilhard had often corresponded, and when he accepted Granger's invitation to fill the position as geologist with the 1930 expedition, one of the most brilliant scientific minds of his time was added to Andrews' staff.

Altogether, the explorers returned to Inner Mongolia with seven Americans and Europeans, twelve Chinese, and eight Mongols; the transport consisted of four Dodge Brothers automobiles and a caravan of fifty camels purchased by Joel Eriksson. On May 18, Young and Liu took the supplies to Kalgan, which was now so hostile to non-Chinese they were stoned in the streets and loudly denounced as "foreign devils." From Kalgan, everything was taken by carts to Joel Eriksson's mission at Hatt-in-Sumu, where it was loaded onto Tserin's waiting camels. To help finance his faltering war against Nanking, General Yen Hsishan had placed numerous tax stations along the railroad from Peking to Kalgan. In addition, tax collectors in Kalgan were extracting squeeze for themselves on the side. Young and Liu had to negotiate eight tax stations en route to Kalgan, and leaving the city required a permit costing $320 in order to use the road. Indeed, between taxes, secret payments to Mr. Sun to influence the Cultural Society in Andrews' favor, and the squeeze he slipped to various other officials, Andrews estimated that he paid about $10,000 to get the expedition into the field.

At eleven o'clock on the morning of May 27, the explorers left Kalgan. By four in the afternoon, they reached the motor inn at Changpeh. Here a detachment of Yen Hsi-shan's soldiers "asked us for every possible passport and tax receipt," Andrews noted, hoping that we might lack some document that would give them an excuse to demand squeeze. Balked at every turn, at last they triumphantly announced that we could not proceed unless we had permits from the Nanking government. Since they were at war with Nanking and would not have recognized the passports anyway, this was too absurd. A considerable crowd had gathered in the courtyard to watch the proceedings, and when I produced our Nanking *huchao* the bystanders roared with laughter. The soldiers lost so much 'face' over the proceedings that they retired precipitously, leaving us in peace."

By the following afternoon, the men were camped at Hatt-in-Sumu. "The tents were [pitched] in the usual place," Andrews gloated, "and we had a very jolly dinner. Granger and I were happy to be in Mongolia again. Every article of our personal equipment was in its usual position in the tent, and we had difficulty believing that we had been away at all." But on May 31, nature abruptly brought them back to reality. While the others were having tea in the mess tent, Thompson spotted a massive yellow cloud looming in the northwestern sky. Within three hours, the temperature dropped 40 degrees, the tents had collapsed under gale-force winds, and the sun disappeared into a curtain of thick dust. Yet this was not the usual Gobi sandstorm. Instead, the air was choked with fine sediment from the region's cultivated fields, which permeated everything including cameras tightly wrapped in triple boxes and played havoc with the automobiles' engines, clogging carburetors, gas lines, gears, and brakes. It took weeks before Mac Young, often working under terrible conditions, was finally able to correct the mechanical problems caused by the storm.

Before taking the entire expedition on the arduous journey to the sand-encircled lake where the nomad, Halchin Hu, had told Eriksson of the existence of rich fossil beds, Andrews, Young, Granger, Teilhard, and a Chinese assistant, with Halchin Hu as their guide, set out to reconnoiter the location. After struggling across miles of nearly impassable terrain—in places, their two cars practically had to be carried through deep sand—they reached the lake, called Tukhum Nor, only to discover that it was a disappointing hoax. The water had dried up, and what few fossils could be seen were badly weathered and of little value. All they found, Andrews reported, was "a shallow depression of sun-baked mud, stained by alkali, and surrounded by a barren waste of sand bathed in streaming heat waves. Our Mongol could hardly believe his eyes. He had seen the lake just after unusually hard summer rains when the depression was filled to overflowing. But that is like most desert lakes—here today and gone tomorrow."

The men plodded across the lake bed to the spot where the humiliated Mongol claimed to have found extensive fossil deposits. Now and then, Halchin Hu proudly indicated bits of bone. "'Here are the *lung-gu* [dragon bones],' he announced, 'many bones. . . .' That gave us pause," Andrews commented. "So this was what we had pushed our way through miles of sand to see! A dry lake of stinking alkali mud and fossils enough to fill a gasoline box. . . . After a [while] we began to see the humor of the situation. . . . It was disappointing, extremely so, but that was about all. Had the entire Expedition wasted time by going there on camels we could not laugh so heartily. As it was we would push ourselves out of the sand and begin again." The daring quest to penetrate the eastern dunes was abruptly abandoned.

After storing extra gasoline and food at Hatt-in-Sumu, the expedition headed north. Roaring past the telegraph station at P'ang Kiang, whose opium-besotted attendant had not left his desolate post for four years, they halted for a night in a ruined temple, then pushed on to the former Mastodon Camp, where bones were still weathering out of the ground in large numbers. Leaving this location after a short reconnaissance, the cars drove eleven miles away to an escarpment overlooking deep red and gray ravines. "It was not a region to be entered without due thought," Andrews recalled. There were no Mongols in the vicinity, the nearest well was miles away, and the ravines were infested with wolves, which prompted the men to name their new home Wolf Camp.

Wolf Camp rewarded the paleontologists lavishly. Almost immediately, Granger and Thompson made a spectacular discovery: the skulls and bones of a mother and infant shovel-tusked *Platybelodon*, both of which had died together only a few feet apart. The next morning, Buckshot found a superb *Platybelodon* jaw, and the other diggers excavated a maze of carnivores mixed in a heterogeneous mass grave that appeared to have once been a bog. Two days later, while studying geological formations south of Wolf Camp, Teilhard de Chardin stumbled upon an amphitheater in the badlands littered with mastodon bones—the site of another bog in which dozens of Pliocene animals had perished. "No one suspected," Andrews said, "that we were about to excavate what will go down in history as one of the world's most remarkable fossil deposits. . . . Twenty [*Platybelodons*], we know, were trapped; probably many more than that. . . . For six weeks the men worked at exhuming gigantic bones and skulls until the hole assumed enormous proportions."

Life at Wolf Camp settled into a delightful routine during the prolonged excavation of the quarry. Every day, Granger, Thompson, Garber, and the Chinese scientists, Young and Chang, took their lunch to the pit, where Buckshot and two assistants had erected their tent to protect the excavation from wandering nomads. Teilhard had ample time to conduct an exhaustive survey of the region's geology. Lieutenant Wyman compiled detailed maps of the topography, and Mac Young worked on the four cars almost daily, restoring them to perfect running condition after the damage they had suffered during sandstorms. Every other day, Andrews and Wyman went out to shoot antelope for the evening meals. Wyman, something of a novice at hunting, thrilled to the chase, but by now Andrews was decidedly jaded. "It is not sport," he lamented, "but we were after the meat. Twenty-five men in camp eat an astonishing amount." As always, the Chinese cooks were masters at preparing antelope, gazelle, deer, and game birds a dozen different ways. And with the addition of rice, dried and canned vegetables, soups, and freshly baked bread—finished off with delicious pies or cakes and coffee—no one could complain about the exceptional quality of the expeditions' meals. "My cooks," Andrews boasted, "have always been nothing short of ingenious, even when our rations were slim."

On June 23, drenching rain and seventy-mile-an-hour winds slammed into the camp. The tents collapsed and the men's sheepskin clothing was soaked in mud. When the storm subsided two days later, Granger was horrified to discover that one of the crates used to anchor the sides of the mess tent in the wind had split open, resulting in the loss of ten precious gallons of alcohol. Unless a new supply of alcohol was brought from Peking, there was no way to thin the shellac used to preserve fossil specimens.

Reluctantly, Andrews and Young started for the city through the sweltering summer heat, driving two cars loaded with eighteen crates of fossils that had already been processed for shipment. After Liu distributed generous squeeze to local officials to allow their entry, Andrews proceeded to Peking, while Young rode with the fossils from Kalgan in an open freight car swarming with pigs and chickens. At almost every stop, the stationmaster demanded from $2 to $10 in squeeze to prevent the boxes from being taken off. A military truck from the American legation met Young at the station in Peking, hastily loaded the crates, and spirited them to the headquarters compound without incident.

Driving back to camp was an exhausting ordeal. Three times in a quarter of a mile the cars became mired in a swamp and sank in mud up to the axles, requiring seven hours to extract them. Next came a steady downpour near Chap Ser, the most dangerous bandit stronghold in Inner Mongolia, which the cars sped past at full throttle despite the slippery road. Twenty miles farther on, they encountered the yurts of the expedition's caravan, where they spent the rest of the night sheltered from the driving rain. Andrews, Young, and Liu arrived back at Wolf Camp on July 14, and after handing out mail and newspapers from Peking, they celebrated a Bastille Day dinner in honor of Teilhard de Chardin.

In the interim, extraordinary paleontological treasures had continued to emerge at Wolf Camp, awaiting the all-important alcohol before they could be prepared for removal. "The great death-trap of shovel-tusked mastodons was almost exhausted," Andrews reported. A dozen jaws and skulls, about thirty tusks, and more skeletal parts had been exposed. In addition, the skull of a peculiar rhinoceros with enlarged nasal bones was unearthed, along with remains of extinct species of deer and foxes. The morning after Andrews and Young returned, Albert Thompson made an astonishing discovery: an adult female mastodon lying on her side. In the pelvic cavity were the skull and jaw of an unborn baby. "A prize if ever there was one," Andrews exclaimed. "Thompson performed the accouchement with Granger in attendance as consulting physician. The rest of us watched and offered gratuitous advice."

Shortly thereafter, more specimens were loaded onto eleven camels for the trip to Hatt-in-Sumu. While in Peking, Andrews had learned that Yen Hsi-shan and Feng Yu-hsiang were getting the worst of their struggle with the Nationalists. If their campaign failed, Andrews feared that hordes of retreating soldiers and deserters would turn to banditry, threatening the railroad out of Kalgan and seeking refuge on the Mongolian plateau. Anticipating it could require months to get his crates to Peking, during which they might be completely lost or confiscated, Andrews decided to begin transporting as many as possible by car. With Young, he departed from Wolf Camp on July 22, taking Teilhard de Chardin, who had completed his geological studies and was returning to Paris. The Cultural Society's representative C. C. Young was also leaving to be replaced by another man, named W. C. Pei. Wyman was likewise being recalled to duty with his unit, having finished his mapping projects.

When Andrews returned from Peking, he brought along his friend Lieutenant-Colonel Margetts from the American legation. During his absence, the scientists had recovered bones of extinct giraffe, four genera of carnivores, bovids, rodents, deer, rhinoceros, and the skulls and jaws of more shovel-tusked mastodons. Together with those obtained from the quarry south of Wolf Camp, the paleontologists had excavated about twenty skulls of *Platybelodon* in every stage of growth from unborn infants to adult bulls. Granger and Andrews both agreed that the area around Wolf Camp, with its dried bogs and inlets along the shores of an ancient lake bed, would be an excellent place to seek fossil primates. But such a quest, they concluded, would require several years of careful searching for the smaller, more fragile bones of primates that may have been trapped along the marshy edges of the lake.

From Wolf Camp, the expedition moved fifty miles west to a rugged escarpment Granger had examined briefly in 1928. The new base of operations was named Camp Margetts in honor of Andrews' guest. Its deeply eroded formations were Eocene and Oligocene in age, and hardly had the scientists examined the area before skeletons of *Baluchitherium* were found, along with titanotheres, the grotesque *Entelodon*, and numerous other species. Among the strangest discoveries were animals belonging to the chalicothere family—an ungulate described by Andrews as "a veritable paradox," with a body and head resembling a horse, but with enormous claws instead of hooves.

On August 13, Young, Andrews, and Colonel Margetts started back to Kalgan with more specimens. Because it now seemed certain that Yen Hsi-shan would be defeated by the Nationalists, it was increasingly urgent to get the crates to Peking as rapidly as possible. Nine days later, Andrews and Young were again headed back to Camp Margetts in a blinding rain to pick up more fossils. On August 30, Andrews drove out of the Gobi for the last time, taking Young, Liu Hsi-ku, and W. C. Pei from the Cultural Society. Osborn had instructed Andrews to open formal negotiations for another season's work in Mongolia, a prospect Andrews knew would involve tedious, if not impossible, efforts to cut through the Cultural Society's lingering resistance. Granger, Thompson, and the rest of the expedition stayed behind to complete the fieldwork, as Camp Margetts turned out to be more rewarding than anyone anticipated.

At most, Granger expected to spend another month in the field. Already, the weather was turning cold, and with the withdrawal of Yen Hsi-shan's troops from the area around Kalgan, bandits were again on the rampage. Near Chap Ser, thirty brigands had killed two Chinese and robbed their cars. A week later, when Young was returning to camp to pick up more fossils, three assailants opened fire at him with pistols just beyond Chap Ser. Young fired back at one of the bandits, smashing a stone close to the attacker's face. Either the bullet's steel jacket or a fragment of rock gouged the bandit's cheek open and tore off half his ear. By then, a dozen bandits had cut loose with rifle and pistol fire. Some had mounted horses and were galloping after him when Young slowed his car, took aim with a rifle held in one hand, and killed a horse, toppling its rider. With that, the brigands abandoned the chase and retreated.

By the end of September, Granger broke camp and the remaining specimens were sent by camel to Hatt-in-Sumu. The staff returned by automobile just as a blizzard struck on October 2. Nor was this the only bad omen. All-out war erupted a few days later between rival gangs of bandits, numbering four or five hundred men.

Remarkably, the expedition's cars were the last to pass unmolested along the road for several months, and all the specimens made it safely back to the Peking headquarters. Spread out on the laboratory floor, Andrews wrote with pride that they were an impressive sight:

It was the largest and one of the most important collections ever taken out of central Asia. The dinosaur eggs and some other specimens discovered in previous years were more spectacular and had more popular interest, but from the standpoint of pure science they hardly surpassed this collection. It gave us additional proof that the central Asian plateau was one of the greatest centers of origin and distribution of animal life during the Age of Reptiles and the Age of Mammals. It showed us that Mongolia was even more favorable for the development of many types of mammals than was Europe or America, and continued to be so, long after these groups had disappeared in other parts of the world. It gave much additional knowledge of the climate, vegetation, and physical conditions of this great incubator of world life.

True, we have not been successful in one objective of our search—the "dawn man." It is a scientific tragedy that Chinese opposition to foreign investigation should end our work just when that goal might be attained. Still, we have shown the way, broken the trail as it were. Later, others will reap a rich harvest. We are more than ever convinced that central Asia was a paleontological Garden of Eden.

CHAPTER 20

Surveying the bountiful yield of their 1930 labors, Andrews and Granger were concerned that the Chinese would trump up some last-minute excuse to renege on their agreement and refuse to hand over the fossils. But their fears never materialized. Despite Andrews' declaration in a letter to Osborn that he "didn't trust Chinese farther than he could throw a mastodon," the Cultural Society's representatives duly inspected the fossils, arrived at an amicable division of duplicate specimens, and released the crates for shipment to New York.

Encouraged by the Cultural Society's unexpectedly cooperative attitude, Andrews and Osborn seriously began to entertain the idea of another expedition to Inner Mongolia in 1931. Nor had they ruled out the alternative plan of combining forces with the Academy of Sciences in Leningrad to explore Russian Turkestan. Everything came to a temporary halt, however, when Osborn's wife, Lucretia, died following a prolonged siege of illness—news that was communicated to Andrews in a brief cable dated August 26, 1930:

> Beloved wife passed away peacefully this morning. She was always devoted to you and your interests.
>
> Henry Fairfield Osborn

In September, Andrews approached the mercurial Chang Chi, the chairman of the Commission for the Preservation of Ancient Objects. Acknowledging that the Museum had assured the commission that the 1930 expedition would be its last, he argued that Inner Mongolia's fossil deposits had proven so lucrative—and its Pliocene lake beds offered such ideal conditions to search for ancient human remains—another season was urgently needed to fully explore the region. Andrews suggested a joint Sino–American Museum expedition, with a Chinese coleader and equal scientific representation from both countries. As usual, the Museum would finance the venture. But Chang Chi's aloof response amounted to nothing more than a vague promise to discuss the matter with the full commission in the spring of 1931.

While Granger and his assistants stayed at the headquarters preparing the latest collection for shipment, Andrews booked space on the Trans-Siberian Railway to Paris, where Yvette had established temporary residence. Renting rooms near Yvette's apartment, he played with George and Kevin, and engaged in "long and painful" conversations with Yvette. But there was no chance of reconciliation, and the details of a divorce were ironed out while Andrews was in Paris. Added to the pressures of his career, certain incompatibilities had arisen over the years—intractable differences in values and outlook. Faced with an ever-widening gulf between Andrews' driving ambition and her own needs, it was inevitable that Yvette should have sought companionship with another man, whose identity Andrews had not yet deduced was Chips Smallwood, one of his closest friends. And there is reason to believe that Yvette was harboring a darker secret: that Andrews was probably not the father of their younger son, Kevin, an issue that would not arise for many years.

Once the legalities were settled, Andrews sailed for the United States aboard the *Mauretania*, arriving in New York on January 2, 1931. Scarcely two months later—on March 30—a French court granted the divorce, allowing Yvette's claim of "desertion." The press immediately began a feeding frenzy regarding the breakup of the glamorous and famous couple, and a number of papers published absurdly sensationalized accounts of the circumstances leading to the divorce.

Along with the monetary strain caused by his divorce, Andrews had suffered heavy financial losses resulting from the stock market crash in October 1929. He could, of course, still rely on his Museum salary of $6,187.50 a year, plus earnings from writing and lectures—sufficient to live comfortably by Depression standards and finance his sons' education in England. However, because of his divorce, market losses, and the uncertainty of political affairs in China, Andrews was rightly concerned about his future. Before his 1930 expedition, he had indicated as much in a letter to Osborn written on May 24. Upon learning that his devoted friend and mentor—then seventy-three years old—was planning to retire as the Museum's guiding spirit in 1932, Andrews was apprehensive regarding the effect Osborn's departure would have on his own position within the Museum's hierarchy:

> While you are president of course I know that I am all right. But were you to leave the presidency before I had returned [from China] for good, what would become of me, so far as the Museum is concerned? I *love* the Museum—I can't think of myself as being associated with any other institution. Still, after the life I have led *just any* job in the Museum wouldn't make me happy. I've got to have . . . a place that will give me a good deal of freedom of action. I couldn't be tied down too much. If I were, I might just as well accept some of the . . . outside administrative positions which I have been offered, at three or four times what the Museum could ever pay me.

Osborn appears to have minimized these concerns by continuing to encourage Andrews to pursue another expedition in 1931, as well as maintaining contacts with Russian scientists. Whatever changes Osborn's retirement might presage, he felt that keeping Andrews in the field during this crucial period outweighed any questions regarding his future position within the Museum.

Following his arrival from Europe, Andrews passed the months from January to May living with Osborn, the publisher George Putnam, and members of the Colgate family, with whom he had remained friends. He buried himself in social events, writing, lecturing, polo, tending to the expeditions' business affairs, and assisting Granger with the enormous collection of fossils that had reached the Museum in the winter in 1931. Much of his time was also consumed with supervising plans for the publication of the monumental summary of the expeditions' achievements entitled *The Natural History of Central Asia*. Already, some ninety-six papers and reports dealing with paleontology, geology, botany, and zoology had appeared in the Museum's *Novitates* and *Bulletin*, along with a flood of popular articles in magazines and newspapers. But apart from these preliminary papers, Andrews had set aside funds from the expeditions' budget to underwrite twelve oversized volumes beginning with his own account—*The New Conquest of Central Asia*—designed to preserve an in-depth record of the expeditions' legacy. Eleven additional volumes written by the project's staff members and a battery of curators at the Museum who had analyzed their discoveries would complete the series, providing a detailed record of everything the explorers had learned about the Gobi. It was a massive undertaking, unique in the annals of modern scientific exploration. Regrettably, the Depression, combined with cost overruns, left only enough money to actually print seven of the intended twelve volumes.

Andrews visited Washington, D.C., in March 1931, where, amid an illustrious gathering of dignitaries, he joined a select few recipients of the National Geographic Society's prestigious Hubbard Medal, an honor previously awarded only to Peary, Bartlett, Shackleton, Amundsen, Byrd, Stefansson, and Lindbergh. In the June 29, 1929, issue of the *New Yorker*, Andrews had been the subject of a profile by Helena Huntington Smith entitled "Hunter of the Snark." And later that summer, he would receive another accolade when the August issue of *Vanity Fair* announced Andrews' nomination to the magazine's annual Hall of Fame.

Early in May, he sailed for China via Vancouver. Hardly had he reached Peking before his hopes for a renewed expedition in 1931 began to fade. Not a single member of the Cultural Society from Chang Chi to the lowest-ranking officials would grant him an interview. Determined to wait out these stonewalling tactics, Andrews settled down in the headquarters compound to work on his *New Conquest of Central Asia:*

> Dr. Sven Hedin . . . happened to be doing a book at the same time and Dr. Davidson Black was studying the remains of the "Peking Man." . . . All three of us found we could work best at night. I played polo in the afternoon, got physically relaxed, and had a light dinner. A pile of sandwiches and a bottle of beer were placed on a tray in my office. As the multiple activities of the day ceased, the compound became quiet and only the calls of street vendors sounded beyond the walls. Then I settled down to write.

Usually about three o'clock in the morning I telephoned Hedin and Black. We would meet at the "Alcazar" or "International," night cafés of somewhat dubious reputation, have scrambled eggs, dance with the Russian girls, and then go home to bed. I woke only in time to bathe, have luncheon about three o'clock, and ride or play polo.

As the weeks passed, no concessions from the Cultural Society were forthcoming. Chang Chi and his underlings flatly refused to meet with Andrews for reasons that were soon clear: a statement eventually was issued by the society declaring that "the American Museum can do no more work in China." In a curt letter, Andrews was informed that the commission was planning its own expedition into Inner Mongolia and Chinese Turkestan. If Andrews and his staff wished to remain in Peking and *study* whatever fossils were brought back by the commission's explorations, they were welcome to do so—a suggestion that infuriated Andrews and Osborn, who saw it as an overt attempt to freely utilize the Museum's scientific expertise. "It is perfectly impossible for the Chinese to do this paleontological work," Osborn fumed. "If they keep us out [of the field], the whole project comes to an end."

Andrews had more harsh words regarding the Cultural Society's refusal to consult with him. "Their attitude toward me has been absolutely insulting," he ranted in a letter to Osborn on June 10, 1931, "and the remarks that I have heard from good sources . . . made by members of the Commission are pretty hard to swallow. They say that the Am. Mus. is one of the greatest robbers that China has ever harbored, that we have stolen the most priceless things from China and a movement is on foot for an official demand that we return to China *all* the collections, fossils, birds, mammals, fish and reptiles that have been collected here in the last ten years!"

Under the circumstances, Andrews wrote to Osborn, "it seems wise to rent the house for a year. . . . Then we will have no expense but will still maintain our base in China. The entire political mess is so uncertain that the whole business might change any day and the wretched Commission [will] have to flee for their lives." After Andrews dismissed most of his loyal servants (the indispensable Lo stayed on as Number

One Boy), the headquarters was rented by a Captain Brown, who was attached to the American legation.

Andrews sailed for Vancouver on the *Empress of Japan*. While en route, he received shocking news that plunged him into despair. His intimate friend, confidant, and one of the expeditions' most valued members, McKenzie "Mac" Young, had been killed under mysterious circumstances. In August, driving from his family's home in New York State to Hollywood (Mac's raffish good looks had landed him a part in a motion picture), he had picked up two young hitchhikers. Somewhere along the road, they stopped to refill a canteen at a spring, and while Mac waited in the car, the men slipped a sedative into the water. When Mac drank from the canteen, he became so drowsy that he pulled off the highway to sleep. He awakened hours later to discover that he had been severely beaten about the head, robbed of his cash, and left in the car.

Apparently, he never fully recovered from the blows to his skull, and even though he continued his westward trip, he complained of crushing headaches. On September 3, his body was discovered in his car on a remote stretch of road near Eureka, California. He had been dead for two days from a gunshot wound to the back of his head. His .38-caliber pistol lay on the seat beside him, along with a pawn ticket for $16.50 for his watch and chain, and a letter to his sister written the morning of his death that brimmed with his usual optimism.

A coroner's inquest ruled Young's death a suicide, dismissing the wound to the *back* of the head, and the case was quickly closed—too swiftly for Andrews' satisfaction. Overcome with grief, Andrews wrote, "I cannot believe that verdict to be true. Unless it were done when [Mac] was temporarily insane from the pain in his head, he never would have taken his own life. Mac was only thirty-seven years old and he had much to live for." Even today, there are aspects of the death that remain unresolved, and suspicions of an incompetent investigation and subsequent cover-up still surround the case.

In August, Osborn set out on a round-the-world journey prior to his retirement, planning to stop in Leningrad to meet with colleagues at the Russian Academy of Sciences. As for the mess in China, everything was at a standstill. Granger had appealed to the State Department for assistance, but the American legation in Peking advised against further contact with the Cultural Society unless political conditions changed drastically. Working on the Museum's behalf, supporters within the Chinese scientific community, including the highly regarded Teilhard de Chardin, were unable to soften the society's stance.

Early in December, Captain Brown moved out of the headquarters compound, and Andrews instructed Lo to find another tenant for six months. With no signs of progress in sight, Andrews resolved to return to Peking in May 1932. He intended to force a showdown with the Cultural Society, but again he ran squarely against Chang Chi. Not only did he refuse to receive Andrews, but he also issued a public denouncement of the Central Asiatic Expeditions' "reprehensible" actions. By now, the Museum's plight, with accusations being hurled back and forth almost daily, made headlines throughout the United States, Europe, and Asia. Some months earlier, the Cultural Society had angrily charged that Andrews' expeditions had gone into the field with only "hunting permits" that prohibited scientific exploration. Incensed by this statement—and wishing to diffuse its impact in the press—Andrews drew up a rebuttal that Osborn distributed to all news agencies. For ten years, Osborn stated emphatically, the Central Asiatic Expeditions' explorations in Outer and Inner Mongolia had revealed extraordinary insights into the scientific legacy of these regions. Working with the consent of both governments, under permits that spelled out exact terms for the expeditions' activities, the Museum had spent over half a million dollars in these countries, and had conformed in every way to local customs and restrictions.

Moreover, the terms of their agreements had been scrupulously upheld. After the "generous assistance" in educational opportunities and the duplicate specimens given to China by the Museum, "one might expect a small degree of gratitude," Osborn declared, adding that the expeditions had never taken one object of commercial value out of the country—though by now the dinosaur eggs could have commanded handsome prices had more been offered for sale. That an attitude of deliberate and prolonged obstruction should have been adopted by the Cultural Society not only toward the Central Asiatic Expeditions but all foreign science, was, in Osborn's words, "nothing less than a scientific tragedy. It is most deplorable that a group of self-appointed individuals . . . should be allowed to bring lasting discredit upon their own country. Their hostile attitude toward foreign science is utterly without precedent in international relations."

As tempers continued to boil, Chang Chi sent Osborn an editorial from the *China Weekly Review* that was highly critical of the Museum's explorations in Inner Mongolia. Osborn, outraged by the article, immediately fired back an openly hostile letter in which he debunked the *China Weekly Review*'s position, commenting that it was "quite at variance with other editorials from foreign and English language publications which have been brought to my attention." This vitriolic war of words, which raged for months with increasing intensity, spelled the end of Andrews' explorations in China.

Realizing that any prospects for further exploration in Chinese territory were impossible, Andrews set about the "unhappy task" of auctioning the furniture and equipment from the expeditions' headquarters in Peking. Suddenly, the once-lavish compound was strangely empty, except for Andrews' personal possessions and art treasures. "I could hardly bear to walk through its rooms," he wrote. "After so many years of feverish activity, interesting visitors, and glittering parties, it was like a tomb."

With all other options closed, Andrews sought to shift his base of operations to Manchuria, now heavily under Japanese influence. Manchuria's strategic location, rich natural resources, and efficient railroads made it an ideal staging area for Japanese infiltration of China. Engineered by pro-Japanese factions, Manchuria's independence from China was declared on February 18, 1932. By March 1, the country had been renamed Manchukuo, and its capital was moved from Mukden to a rather dingy industrial city known as Changchung.

In the meantime, the former Chinese emperor, Pu Yi, his wife, Elizabeth, several concubines, and a sizable number of retainers had been living in the Japanese concession in Tientsin since fleeing the confines of the Forbidden City in 1924. While in Tientsin, Pu Yi had lived as a closely guarded puppet, surrounded by Japanese agents, attending to largely meaningless "affairs of state," languishing in nightclubs and cafés while Elizabeth sank into opium addiction out of boredom and loathing for her weak-willed husband. All evidence suggests that Pu Yi decided to cast his lot with the Japanese because of their promises to restore him to the Manchu throne. After reviving the emperor's "power" over Manchukuo, Japan held out the promise of granting him "legitimate" control throughout China, a cynical ploy that obscured Pu Yi's ability to see how blatantly he was being manipulated. On March 9, Pu Yi was formally named Manchukuo's "Chief Executive," ensconced in a hideous Russian-built compound in Changchung, and placed under the watchful eyes of the Japanese.

After taking a four-month lease on a sprawling, fully staffed palace in Mukden belonging to a once-powerful nobleman, Prince Chi, Andrews visited Pu Yi in the spring of 1932. Hoping for permission to enter the desert through Manchukuo, he appealed to Pu Yi for help. Although Andrews was cordially received, Pu Yi was powerless to assist with his plans. But a proposal sent to the Japanese vice-minister for foreign affairs, Chuichi Ohashi, proved more successful. Perhaps out of a genuine interest in Andrews' explorations, or a wish to infuriate the Chinese, Mr. Ohashi took up the request with Manchukuo's cabinet. After debating the matter, Andrews was granted permission to launch an expedition into the northeastern Gobi from Manchukuo. As the *New York Times* noted, "The Manchukuo Government's arrangement with Mr. Andrews strengthened a belief that Japan intends to extend the borders of Manchukuo, as well as to occupy [a] considerable area in Inner Mongolia, since the region Mr. Andrews desires to explore lies there." One condition, however, was forcefully stipulated by the foreign minister. "It has to be pointed out," Ohashi wrote to Andrews, "that any Chinese or foreigner who harbors sinister designs upon the new state can not be welcomed and that some Manchurians or Japanese who are well versed in the line of your investigations may be attached to [your] party." Andrews was elated. It was agreed that a headquarters would be established in Mukden, and that specimens would be shipped to New York from the port of Dairen on Manchukuo's southern tip.

As expected, the announcement of the Museum alliance with Manchukuo touched off an angry response from the Chinese, who viewed any cooperation with the Japanese as treachery. At that point, they insisted, only Japan and the American Museum recognized the validity of Manchukuo as a sovereign state. Several Chinese newspapers rushed to denounce Andrews as "arrogant," and his willingness to acknowledge Manchukuo as an independent government cast him in the role of a diplomatic representative of the United States rather than an explorer. In short, they charged, Andrews had confused politics with science. Even worse, the Chinese insisted, Andrews acted entirely on his own, with no authority from the Museum. Osborn was quick to respond to this allegation. As Vice Director of the Museum for Asiatic Exploration, Andrews had every right, Osborn said, to approach Manchukuo—or any other country—after his negotiations with China collapsed. Indeed, Osborn proclaimed to the press, "he had ordered Andrews to transfer his operations to Mukden." As a result of the new agreement, Andrews added with obvious satisfaction, China "will be deprived of the benefits of a great expedition. . . . Instead of getting the most valuable collections without expense, China will get nothing. Manchukuo's cordial attitude toward foreign scientists is in distinct and pleasing contrast with that of Nanking, which has placed all possible obstacles in the path of scientific exploration."

But Andrews' optimism was short-lived. Unwilling to delay its military ambitions for science, the Japanese army began amassing troops along the frontier of Inner Mongolia in preparation for a major thrust into the region. Skirmishes broke out with Chinese garrisons, leaving the area in danger of a wider conflict, which was not long in coming. On July 25, 1932, a letter from Ohashi left no doubt in Andrews' mind that his plans were permanently dashed. While its language was elusive, the implications were unmistakable:

Dear Sir:
    On behalf of the Government of Manchukuo, I take the pleasure of informing you that this government, in consideration of the meritorious achievements of your organization in the past, is ready to render on certain conditions our whole-hearted support to your great scientific work. However, I am sorry to state that the peace and order of the districts where you desire to carry out your excavation is not yet satisfactory and that under the circumstances we can hardly agree that you start the expedition so soon as next Spring, due to the danger contingent upon

your expedition and also to an undesirable effect which the presence of your party may give to uneasy minds of the people.

Therefore I take the liberty to request that you may postpone your plan until the general conditions in the districts sufficiently insure the success of your expedition.

Sincerely yours,
Chuichi Ohashi
Vice-Minister for Foreign Affairs

Finally convinced that the Central Asiatic Expeditions were ended, Andrews returned to Peking in August. He crated his belongings and sent them to Tientsin for shipment to America, bade farewell to his grieving Number One Boy Lo, and closed the headquarters at Number 2 Kung Hsien Hutung. On August 25, Andrews walked through its great red doors for the last time.

At age forty-eight, he had no idea what the future held. He was certain only that nothing would ever again equal the lure of the Gobi or its unknown secrets still lying beneath the sand.

CHAPTER 21

Arriving back in New York in October 1932, Andrews installed himself in a penthouse atop the Hotel des Artistes, a Gothic-style building at 1 West Sixty-seventh Street, just off Central Park West. He converted it into an elegant setting for his collection of antique Chinese furniture and art. From the flower-filled terrace, he could look across Central Park to Fifty-ninth Street and the Plaza Hotel. "The view was lovely in the daytime," he wrote, "and dazzling at night. . . . I made it into a perfect . . . Chinese home." So unusual was the apartment, in fact, that it was featured in the June 1933 issue of *House and Garden*.

Begun in 1915, the Hotel des Artistes was never actually a hotel but rather a complex of artists' studios. It was one of several buildings on the north side of West Sixty-seventh Street originally occupied by, among other celebrities, Al Jolson, Rudolph Valentino, Noël Coward, Isadora Duncan, Ben Hecht, Alexander Woollcott, Edna Ferber, Paul Whiteman, Marcel Duchamp, William Powell, Childe Hassam, and Norman Rockwell. In 1932, the famous illustrator Howard Chandler Christy, a resident of the Hotel des Artistes who befriended Andrews, was commissioned to decorate the Café des Artistes, located on the building's ground floor. Renowned for his depictions of women—the so-called "Christy Girls"—he covered the café's walls with a montage of thirty-six ravishingly seductive nudes in a jungle setting; they still adorn what has become one of New York's landmark restaurants. While Christy was working on the murals, a stream of exquisite models frequented his studio, whom the artist, a notorious womanizer, often took to bed and whose feminine charms he generously shared with Andrews.

Following his return to New York, the Museum appointed Andrews associate director. As he had done earlier, however, when he was made Vice Director for Asiatic Exploration and Research, Andrews insisted on a grander title. On January 5, 1933, he wrote to Osborn (who became honorary president upon his retirement in 1932):

> ... Regarding the title of the new administrative post for which you have designated me, I feel that Associate Director is by no means as strong a title as that of Vice-Director.
> Associate Director rather suggests the position of someone who is associated with an administrative officer for special reasons, whereas Vice-Director indicates in the mind of the public someone who is next in authority.
> I do not see any reason why I should not be designated as Vice-Director in Charge of Scientific Administration. . . . I do feel . . . that in the impression it conveys the title of Associate Director is distinctly inferior to that of Vice-Director.

Osborn arranged for the Museum's director, George Sherwood, and the new president of the board of trustees, F. Trubee Davison, to comply with Andrews' request. Yet it was a hollow victory: Andrews was racked with a discontent that nothing could dispel. Sherwood, he observed, "one of my oldest and dearest friends, was the sort of man who must do everything for himself. . . . With no room for innovation, [there] was nothing important for me to do," Andrews complained in *Under a Lucky Star.* "For the first time in my life I had no objective; no goal. I was definitely unhappy." Andrews, of course, had legions of friends in New York, and his engagement book was always full weeks in advance with parties, dinners, concerts, receptions, and lectures. But this frenetic socializing failed to mask his restlessness and longing for the Gobi.

The prospect of a new adventure suddenly arose with a telephone call from Douglas Fairbanks, Sr., who had visited Andrews in Peking in 1924 after an unsuccessful hunt for a long-haired Manchurian tiger. Having just arrived in New York, he invited Andrews to meet him at the River Club. Ever since his trip to China, Fairbanks explained excitedly, he had wanted to produce an epic film about the country, but he was unable to find a suitable script. Only recently he had devised a story with the help of Madame Wellington Koo, wife of the former ambassador from China to the United States. Fairbanks had given the synopsis to a screenwriter and was thrilled with the finished script. "Everything centers about a young Chinese airman who saves the country in the midst of a war," he told Andrews. "Of course, we won't say so but it's war with Japan. You and I know that's bound to come soon."

Although Fairbanks planned to shoot most of the picture in Hollywood, some of the scenes would be filmed in China. Because of Andrews' knowledge of the country, Fairbanks offered to make him "assistant producer or something of that sort," at a salary of $1,000 a week. "If that isn't all right," he said, "tell me what you want. Will you do it?" Andrews jumped at the offer. Fairbanks confessed that he also suffered from an insatiable wanderlust, and during their conversation, Andrews happened to mention that for the past quarter of a century he had not spent an entire year in any one country. Fairbanks instantly leaped up and called to his wife in the next room. The bedroom door opened and the legendary screen idol Mary Pickford, "America's Sweetheart," appeared. "We were introduced," Andrews recalled, "and Fairbanks promptly announced: 'Roy says he hasn't spent twelve months in any country for twenty-five years. What do you think of that?'

"'I think,' she said, smiling wryly, 'that he will be a bad influence on you. I don't know him but I definitely do not approve.'"

Unfortunately, the film project never came to pass. In spite of a flood of lengthy telegrams regarding preliminary arrangements, Fairbanks had too many prior commitments, and when his son fell seriously ill in Paris, he sailed for an extended stay in France. Andrews went so far as to sketch out a scenario for a motion picture on the life of Genghis Khan and another based on his memoir, *Ends of the Earth*, which had appeared in 1929. But Fairbanks, although well intentioned, could not settle down long enough to develop any of these ideas—much to Andrews' disappointment.

His spirits were temporarily lifted when the cover of the *New York Times Book Review* for February 26, 1933, announced the publication of his long-awaited masterpiece, *The New Conquest of Central Asia*. Under a banner headline, WITH ANDREWS IN MONGOLIA, the critic Henry E. Armstrong praised the book as an "absorbing chronicle," and extolled the depth and liveliness of its narrative style. Other papers across the country followed suit, and *The New Conquest* was hailed as a triumph, an enthralling blend of science and adventure that remains, in

Stephen Jay Gould's words, "one of the most captivating narratives of exploration ever written." Apart from Andrews' text and numerous photographs, the volume also contained sections by Walter Granger, Clifford Pope, Nels Nelson, Charles Berkey, Frederick Morris, Ralph Chaney, A. W. Grabau, L. Erskine Spock, and Osborn. In conclusion, there were commentaries on unresolved problems still awaiting further research in all of the disciplines studied while in the Gobi. They were written by members of the expeditions' scientific staff and three specialists, the mammalogist Glover Allen, the ichthyologist J. T. Nichols, and the invertebrate paleontologist Chester A. Reeds, whose work on the collections from Mongolia and China was confined to the laboratory.

Andrews immersed himself in his duties as president of the Explorers Club, a post he held from 1931 to 1934, until he was succeeded by Walter Granger. Incorporated in 1905, the club boasted an exclusive roster of the world's most famous explorers, and its highest honor, the Explorers Club Medal, was awarded to Andrews in 1932. Each Tuesday, he attended luncheons at the Dutch Treat Club, a favorite gathering place for publishers, writers, and musicians. He also belonged to the Century Club, the exclusive haunt for some of the city's most prominent figures. Andrews occasionally played in what can only be described as "high society" baseball games, which pitted a team of celebrities organized by Lowell Thomas against another led by Theodore Roosevelt, Jr.

Still unable to settle down to routine work at the Museum, Andrews decided to revive the expedition to Russian Turkestan that he and Osborn had been toying with since 1929. In the summer of 1933, Andrews sailed for Europe aboard the *Berengaria*. After touring the French château country, Yugoslavia, and the Austrian Tyrol, he ended up in Berlin. Getting drunk in a beer hall one night, Andrews squarely faced the source of his misery. "You are an explorer," he told himself, if his account in *Under a Lucky Star* can be taken literally. ". . . You can't be happy in the conventional life of a city. It doesn't fit. Go back to the desert where you belong." By three o'clock in the morning, Andrews tells us, he was mentally at peace. He had resolved to throw himself wholeheartedly into organizing a joint expedition under the aegis of the Museum and the Soviet Academy of Sciences. "Russian Turkestan in the summer. Iran in the winter. I'd stay for five years," he envisioned in an alcoholic haze. "Never come out. To hell with cities."

The next day, Andrews flew to Leningrad for a cordial meeting with A. Borissiak at the Academy of Sciences. He then went to Moscow for discussions with other colleagues and Communist officials, including the assistant commissioner for foreign affairs, a man named Karakhan, whom Andrews had known in Peking, where he served as Russia's ambassador. The terms of the proposed expedition were agreed upon with relative ease. Andrews would supply the money and some of the scientific staff; the Russians were to provide equipment, transportation, and additional scientists. The collections would be shared equally.

Andrews was confident he could count on the support of Osborn's successor, Trubee Davison, a wealthy businessman who had served as Herbert Hoover's assistant secretary of war for air. While he was not a scientist, Davison was a competent administrator and well connected in New York's financial circles. Both Davison and Sherwood gave their blessing to the idea of a joint expedition with the Russians. Regardless of the Depression's grip on possible financial resources, Andrews optimistically laid plans for a fund-raising campaign to underwrite the expedition. Within a few weeks, he inaugurated a series of lectures and radio programs on science and exploration that raised the first $26,000 toward his new venture.

Nevertheless, the Russian project soon collapsed. Andrews received an about-face letter from William Morden—the zoologist-explorer with numerous Russian contacts who had first suggested the idea—warning that the Russians' promises were not to be trusted. "They are," Morden declared, "without exception the greatest, most enthusiastic, long-distance liars the world has ever known. . . . You will be told that you will have the assistance of the government, and you will have—and how! They will organize, equip, and run your show for you—and you will pay for it, and how you will pay." Morden also said that nothing could be accomplished without the personal sanction of Stalin and his ministers, a virtually impossible obstacle to overcome.

At this point, an unexpected event diverted Andrews from his concern over Morden's discouraging letter. George Sherwood, the Museum's director, was found one day by his secretary slumped in his chair gasping for breath. He had suffered a heart attack, and after being rushed to the hospital for treatment, his doctors prescribed six months of complete rest. With no one to take his place, Trubee Davison pleaded with Andrews to postpone further attempts to organize the Russian trip, and assume the post of acting director while Sherwood recuperated. For Andrews, still hoping to unravel the complications surrounding the Russian expedition, the timing could not have been worse. Yet he felt obligated to accommodate Davison and accept Sherwood's position for six months.

"I enjoyed myself enormously that summer," he wrote, "for there was a job of reorganization [at the Museum] that needed to be done. Trubee and I worked together perfectly because our minds functioned in the same way. He had complete confidence in me and I not only admired his ability but developed a deep affection for him." Even so, the long hours at the Museum were confining, which began to take a toll on Andrews physically. He found time to ride horseback in Central Park every day, and he bought a string of ponies and played polo twice a week at the Fairfield Hunt Club in Connecticut. But one day in August, he took a terrible spill during a polo match and ruptured two lumbar vertebrae. Another fall, his doctors concluded, could either paralyze or kill him: Andrews would never again mount a horse. "[It] was like a sentence of imprisonment to me," he lamented, "for my athletic interests had centered about horses for twenty years—polo, fox hunting, and steeplechasing. I didn't know what to do."

At the end of six months, it was evident that Sherwood's heart condition was too serious for him to return to the Museum. The trustees asked Andrews to become the full-time director. "I didn't want the job," he asserted, knowing it would entail the kind of administrative responsibilities he had always dreaded. "My heart lay in the desert and I hoped soon to go back to the sunsets and the sandstorms." Davison assured Andrews that he could continue his exploration and that the trustees wanted him to resume fieldwork. "I still hesitated," Andrews said. "That I should become Director of the institution where I had spent all my life was a logical conclusion but I could not be certain it was best for the Museum or for me."

While Andrews was wrestling with this decision, one of his friends, Dr. William H. Holden, unwittingly solved the quandary. He telephoned and announced abruptly, "The most beautiful girl I've ever looked at is coming to my office this morning. Why don't you come and we'll have lunch together?" Andrews needed no further coaxing. He arranged luncheon at the Museum, followed by a two-hour tour of the exhibition halls and preparation areas. When the lady in question appeared at Holden's office, she was even more stunning than Andrews had been led to expect—tall and slender, with dazzling gray eyes and a crown of golden hair, swathed in a mink coat with a fur hat pulled alluringly over her forehead. Her unlikely name was Wilhelmina Christmas—her maiden name had been Anderson—and she had grown up in Chattanooga, Tennessee, and New York City before attending school in the South of France. Attracted to the theater at an early age, she had married an actor of mediocre talents. She left him soon afterward to become a John Robert Powers model before landing a part in Flo Ziegfeld's last Broadway show, entitled *Hot Cha!* In 1929, she married a wealthy businessman, Franklin Christmas, and moved into a fashionable town house at 27 Beekman Place. A few months after their wedding, Christmas entered the hospital for a routine appendectomy and died from a blood clot. At age twenty-two, the glamorous Wilhelmina Anderson Christmas—"Billie" to her friends—inherited a substantial fortune in investments and insurance.

Writing in *Under a Lucky Star*, Andrews recalled his reaction to meeting Billie:

> I couldn't get her out of my mind so I asked her to dine with me Saturday evening. We talked from seven o'clock until two in the morning. . . . I begged for the next evening, but she couldn't so it had to be Monday. All day Sunday I was restless as a fish out of water. I telephoned her twice but her maid said she wasn't home. Monday night finally came. Over cocktails I said, "You are going to marry me. I don't know when, but it's inevitable. You might as well make up your mind to it."
>
> On Tuesday, Trubee asked me again if I would not take the Directorship of the Museum. "Certainly, I'd love to. I am going to get married." He gasped. "But only last week you told me you would never get married again!"
>
> "Of course I did, but that was last week. This week it's different."
>
> He only shook his head. "You work too fast for me," said he.
> It was as great a surprise to me as to him. I had built up a

philosophy of life in which marriage never again would figure. It was all very logical and clear until I met Billie; then the theory collapsed of its own weight.

Three months later, they were married on February 21, 1935. The wedding was held at an undisclosed location and a day earlier than had been announced in the papers due to a mysterious bomb threat. (Andrews suspected it came from someone who disapproved of his stance on evolution.) The couple sailed immediately for the Pacific Coast by way of the Panama Canal. Arriving in Los Angeles, Andrews called Will Rogers; they had barely missed each other in Peking eleven years earlier, but had kept up a desultory correspondence. Rogers invited the newlyweds to lunch at his studio with the actors John Boles and Conrad Nagel. Afterward, joined by Andrews' friend, the journalist and screenwriter Irvin Cobb, they visited Rogers' home and drove to Uplifter's Ranch, where Rogers played polo.

Returning to New York after two months, Andrews and Billie settled into a sumptuous apartment in the former mansion of the late publisher Joseph Pulitzer at 11 East Seventy-third Street. Located on the top floor, the two-story apartment boasted a wide terrace and a distant view of Fifth Avenue and Central Park. Andrews and Billie converted it into a virtual palace. With Chinese-red walls and gold ceilings, they filled the enormous living room with Andrews' Oriental art: antique chairs and tables, cabinets inlaid with mother-of-pearl, delicately painted screens, embroideries, lacquered vases, carved jade, and ceramics—nearly everything having once graced the temples of the Forbidden City before Pu Yi's rebellious eunuchs looted his treasuries. Here, Andrews and Billie would live for the next seven years, an ideal haven from the Museum's Depression-racked turmoil and the scene of countless social events. In time, their friends came to include Helen Hayes, Charles Lindbergh, Doris Duke, Frank Buck (famous as an animal collector for zoos and circuses), Richard Byrd, Barbara Hutton, Amelia Earhart, who was married to Andrews' friend and publisher George Palmer Putnam, Erskine Caldwell, Vincent Sheean, Elsa Maxwell, Bernard Gimbel, the stage and film actress Anna May Wong, Lowell Thomas, and other luminaries from all walks of life. In spite of the prospect that he would never be able to resume his explorations,

"My friends said they thought I was very lucky," Andrews wrote. "I knew damned well I was."

On November 6, 1935, Henry Fairfield Osborn died after a siege of gradually failing health. If any single individual was responsible for Andrews' extraordinary career, it was surely his friend, mentor, and staunch supporter Osborn. Understandably, his passing left Andrews profoundly bereaved. In a memorial published in *Natural History*, he penned the following tribute:

> When I think of Professor Osborn it is always first as a devoted friend and a perfect companion. Perhaps no one, not even a member of his family, knew him in those respects better than I. Since 1923, during the intervals between expeditions, I have lived at his home. Every morning we breakfasted together; at night we talked in front of the fire; often in the days we took long walks through the woods at Castle Rock or in Central Park.
>
> He knew even the most intimate episodes in my life. I brought to him all my problems and he made them his own. Out of the fullness of his wisdom and experience he gave me the advice that directed every phase of my life and work. As I see the last fifteen years in perspective I realize what I owe to him....

Osborn's death occurred just over two months after the loss of Andrews' mother, who had passed away on August 15 in Beloit. (His father would die in Florida in 1939.) Thus his tenure as director of the American Museum began on a sad note, exacerbated by the drastic erosion of the Museum's activities caused by the Depression. Andrews had been installed as director largely because the trustees believed his adventurous image—still the object of widespread public adulation—and his genius for raising money could be used to attract desperately needed financial support. When Andrews accepted the directorship in 1935, at age fifty-one, he radiated a distinguished bearing akin to Osborn's. Always dapper in tailor-made suits, wearing a pince-nez (the result of damage to his eyesight inflicted by the Gobi's glaring sunlight), and never without a walking stick, Andrews exuded the unwavering self-confidence he had always demonstrated in crises. If anyone could lift the Museum out of its financial doldrums, Andrews seemed the logical choice.

To relieve the stress of his responsibilities, Andrews took Billie on an excursion to the Bahamas to visit his friend William Beebe, the zoologist and renowned undersea explorer. Eager to throw herself into new adventures, Billie, equipped with a diving helmet and air hose, could scarcely wait to climb down the thirty-five-foot ladder attached to Beebe's boat into the eerie underwater world of multicolored coral, sponges, and exotic fish. She found the experience "exhilarating," while Andrews, who had never feared anything in nature (except snakes), described the ocean's depths as "terrifying."

Several months after returning from the Bahamas, Andrews and Billie sailed to China on the *Empress of Japan*. In spite of deep-seated reservations concerning the emotional impact of revisiting his past, Andrews wanted Billie to glimpse something of his former life. By cable, he rented a charming, flower-covered house on the "Street of the Small Market" in Peking, and they departed from Vancouver on April 14, 1936. They spent two months in Japan, then traveled from Kobe to Tientsin. As was typical of the spring weather, a dust storm ushered them through Peking's Tartar walls, where some of Andrews' former servants were waiting with flowers. When the gate to their rented compound opened, "the clink of ice in a cocktail shaker sounded," Andrews recalled. "Heavenly music! . . . [And] a diminutive *amah* had a bath . . . drawn for Billie. It was a typical Peking reception."

> . . . Out in the city [he wrote] my fear of coming back ended. I was in Peking again, the same sights and pungent smells, the same street calls, the same pigeons with whistles on their tails circling over the yellow roofs of the Forbidden City! Everything was there but I seemed to have known it only in a previous existence. My house on the Bowstring Street was sadly in decay. No one had lived in it since I left. As we roamed through the courtyards and the vast rooms, thick with dust, I thought of myself in the past, in the third person. I was like an embodied ghost. The past was as impersonal as though I had been a character in history.

With a busy schedule of sight-seeing, visiting Andrews' friends, and exploring out-of-the-way places, they crowded as much as possible into their two-month stay. Still, Andrews felt an impending sense of doom hanging over the city. Notwithstanding diplomatic protests, he observed that "the Japanese were pouring troops and munitions into China. Their swaggering soldiers pushed foreigners off the sidewalks and spewed insulting epithets when a white girl passed."

As Andrews' tenure as the Museum's director wore on, he found himself spending more of his time grappling with ever-diminishing financial resources. One of the worst effects of the situation from Andrews' viewpoint was the end of the Museum's "Golden Age" of exploration, launched by Osborn not long after he was elected president of the board of trustees in 1908 and crowned in the 1920s by the Central Asiatic Expeditions. Even had there been no political complications to overcome, money was no longer available to conduct large-scale explorations in distant parts of the world. Within the Museum itself, however, three important projects were completed.

In October of 1935, the Hayden Planetarium was opened and quickly became a popular attraction. One of Trubee Davison's priorities, he had persuaded the Reconstruction Finance Corporation to underwrite the building; its optical instruments, designed in Germany to recreate the astronomical configurations of the night skies, were donated by the investment banker Charles Hayden at a cost of $150,000. Plans were also under way for a Hall of North American Mammals, which was constructed with almost $250,000 in private contributions and $15,000, a small but crucial amount, squeezed out of the City of New York's strained coffers by Andrews after he deftly overcame Mayor Fiorello La Guardia's opposition to the expenditure. By far the most spectacular of the new additions, however, was the completion of the Akeley Memorial Hall of African Mammals. Although Akeley died in the Belgian Congo in 1926, he had already planned every detail of the hall in a huge scale model. After years of painstaking labor, this extraordinary exhibit, supervised by Andrews' longtime friend James L. Clark, opened to the public in 1936, generating an outpouring of accolades. "Nothing like it exists in all the world," Andrews proclaimed. "It *is* Africa. Not only the animals but the trees, the leaves and grass, and the very earth itself, were brought from the place where each [diorama] was collected. As a rule it is futile to say a thing can never be done better. But I am perfectly willing to make that statement about the African Hall."

By now, the Museum had become second only to the Empire State

Building as a lure for prominent visitors to the city, and among the many luminaries who came to view the Museum's acclaimed exhibits and meet its famous explorer-director were Prince Chichibu, the brother of Japan's emperor, Hirohito, the Duke of Windsor, and the Crown Prince of Sweden.

Yet none of this helped to overcome the Museum's financial pinch. "Drastic curtailments were necessary," Andrews lamented. "It got to the point where, with a million and a half dollar budget, I was trying to save fifty dollars on paper drinking cups and towels; where when a collection worth ten thousand dollars was presented to us, we couldn't take it because we didn't have three hundred dollars to pay the cost of packing and transportation. Appropriation for scientific publication was cut in half [preventing the printing of five of the projected twelve volumes of Andrews' *Natural History of Central Asia* series]. I had to write letters of dismissal or retirement to many of my oldest friends." Even Andrews' salary was frozen at $15,000 a year.

Now, as well, Andrews found himself the target of harsh barbs hurled by detractors. Andrews was viewed by many as an ineffective administrator, a misplaced adventurer whose abilities in the field were unquestionably brilliant but who pursued ill-advised policies behind the director's desk. Just as Andrews himself had feared from the outset, it was evident that his temperament was not well suited to the job, although it is difficult to imagine that *anyone* could have managed substantially better during the Depression's darkest days.

Other critics within the curatorial ranks regarded their flamboyant, high-living director as a self-serving "headline seeker" who used his widely publicized exploits to glorify himself. Such grumblings had been heard as far back as Andrews' 1912 Korean expedition, and doubtlessly reflected professional resentment engendered by Andrews' highly touted achievements—but mixed with a certain element of truth. Along with his oversized ego, Andrews never missed an opportunity to enlarge his status as a celebrity. He was a clever showman with an uncanny ability to draw the public into the excitement of his escapades, something he felt was too often missing in scientific endeavors. He frankly relished being in the limelight, and was a master at manipulating the press as an essential part of his fund-raising skills.

Apparently unwilling to separate his accomplishments from their disdain for Andrews, few of his detractors ever mentioned the tons of paleontological material or the thousands of zoological and botanical specimens added to the Museum's storehouse of treasures by the Central Asiatic Expeditions—or the fact that Andrews was meticulous in crediting his colleagues in the field for their discoveries, or the importance of their laboratory research and publications.

By 1937, a Communist cell within the Museum had joined the chorus of disapproval against Andrews. In the May 1 issue of their in-house bulletin, appropriately entitled the *Red Fossil*, an anonymous letter to the editor appeared under the rubric "Coxcomb Roy":

> You may not know that one of the most delightful effects of the RED FOSSIL is that the coxcomb Director of the American Museum of Natural History, the playboy Roy Chapman Andrews, is exceedingly aggravated because he has had no notice— not even abuse or a mild left-handed compliment. The one thing the poor dear can't stand is being overlooked. His great craving in life is to appear important, especially among the plutocrats of the board, who have at times permitted him to share their crumbs.
>
> He grabs each number of your sprightly journal in search of a slam which he can proudly show to his "Trubee" or mail to J. P. Morgan. He is enraged at the absence of references to himself, when the plutocrats whom he would like to imitate get so much pummeling. . . . Next time pass along a sly inference that the director is not of sufficient importance to rank a reference, since he never had an original idea of his own. . . .

During the Cold War, the Communist regimes in China and Mongolia branded Andrews, along with other American and European explorers, as a criminal and an instrument of capitalism. In 1959, when the author Harrison E. Salisbury visited the State Museum in Ulan Bator, he reported seeing "mug shots," as he called them, of two Americans labeled as "espionage agents" prominently displayed on a wall: they were Andrews and his friend the writer and Asian scholar Owen Lattimore.*

---

*One of the ironies of the collapse of the Communist regime in the Mongolian People's Republic in 1990 has been an extraordinary resurgence of interest in Andrews' explorations, and the droves of tourists that now swarm over the Flaming Cliffs and other landmarks associated with Andrews' legacy.

Most recently, Andrews' reputation has suffered from the "political correctness" mania. Various magazine writers and the authors of a catalog for a student-organized exhibition in 1991 on Andrews' career, shown at the Logan Museum at Beloit College, have lambasted him for his supposedly "patronizing" attitude toward the Mongols and Chinese. In an unrealistic effort to push history forward to reflect today's often naive approach to such issues, Andrews has been denounced for referring to his servants and workmen as "boys" or "coolies." Yet in Andrews' day, these terms had been used by foreigners and Chinese alike for generations. Moreover, to anyone familiar with the Museum's archives pertaining to Andrews' expeditions, it is clear that he regarded many Mongols and Chinese with genuine affection—Badmajapoff, for example, Merin, Buckshot, Liu, Tserin, and numerous scholars and officials with whom he worked closely for years. Upon leaving China in 1932, he found employment for his Number One Boy, Lo, and continued to stay in touch with Lo and his son until the Japanese invaded China in 1937.

Like so many such ventures by colonial powers, the Central Asiatic Expeditions carried indisputable "imperialistic" overtones. Osborn, Andrews, Granger, and everyone else connected with the project never really questioned the Museum's right to enter Mongolia and China in the name of science and carry off discoveries to enhance the American Museum's collections. Even though there were almost no enforceable laws in these countries governing the removal of scientific material in the 1920s, Andrews always negotiated legitimate permits with the governments involved. Yet he was obviously caught in a difficult dilemma. His epic explorations opened the way for much of the modern scientific research in these regions, but with the advent of a Communist government in Mongolia, the approach of all-out war in China, rampant nationalism, and a fierce determination to purge the last vestiges of imperialism—clearly represented by the massive, heavily financed Central Asiatic Expeditions—Andrews found his venture swept away by a tidal wave of history.

CHAPTER 22

Regardless of Andrews' personal and financial tribulations within the Museum, his public adulation remained undiminished. He was in constant demand as a lecturer and author, and was frequently asked to lend his name to a variety of charitable causes, such as the China Relief Fund, whose chairman was Pearl Buck, author of the best-selling novel *The Good Earth*. Socially prominent hostesses competed to entertain Andrews and Billie at their soirees. Autograph seekers hounded Andrews, often forcing him to avoid crowds in public places. And he was regularly sought after for commercial endorsements and magazine advertisements.

Andrews again made headlines in 1937 when he was awarded the Loczy Medal by the Hungarian Geographical Society, and he became only the third American to win the Vage Gold Medal of the Royal Swedish Geographical Society. In addition to his two honorary doctorates and growing list of medals and citations, Andrews eventually was elected to numerous prestigious organizations, including the American Geographical Society, the New York Zoological Society, the American Association for the Advancement of Science, the National Geographic Society, the American Philosophical Society, the California Academy of Sciences, and the Biological Society of Washington.

During his tenure at the Museum, Andrews participated in a weekly radio program. In 1937, the National Broadcasting Company recruited him, together with Admiral Byrd, Theodore Roosevelt, Jr., Lowell Thomas, and the explorer-author Felix Reisenberg, for a series based on their experiences in distant corners of the earth called *Order of Adventures*. Intrigued by the potential of radio, Andrews later appeared on a program every Wednesday known as *New Horizons* on Columbia's *School of the Air*, which lasted for five years.

This exposure intensified his public following to an even greater degree, and triggered a renewed deluge of fan mail. Much of it came from women admirers, some of whom were astonishingly blunt. One of the least subtle letters proclaimed:

Dear "Doc,"

I think you are simply swell. I've just read in the Times you've been decorated. I wish I could decorate you on both cheeks the way the French do, but if I ever put my arms around you I'd never want to take them away—

You are a grand person, one of the world's swankiest and My Gawd, you can park your shoes next to mine anytime you feel the urge. And that's that!!!!

Just a fan.

One letter in particular outclassed all others. Written by a wildly enamored young woman on scented stationery in a large flourishing script, she told of her "warm, mad, eighteen year old heart [beating] furiously" when she wrote Andrews' name; how she trembled when she caught sight of him; that her longing to know him was more than she could endure. She confessed to collecting "all your pictures . . . in a scrapbook; all your books . . . and magazine stories; all the press notices that concern you and one very special distinguished photograph, which I stole from a friend of yours and keep with me always beside me on the pillow. It is the last face I see before the night light is extinguished and the first in the morning when day wrenches the dream of you from me. . . . You are," she continued, "so witty, so wise, so worldly. I think to be loved by you would be to know and experience the essence of the love of all men in one being." She pleaded with Andrews to meet her for afternoon tea at Pierre's. "I shall be in the reception room wearing a Skipper Blue costume with silver fox furs and three rose orchids on my left shoulder. I am tall, blonde, and slim, with skipper blue eyes. . . . How much greater could [my] ecstasy be if I had the intense joy of being near you. . . . I am well nigh fainting with the imagery."

No one knows if Andrews responded to the invitation. The letter was found tucked away among a box of private papers in the possession of his son George. One thing *is* certain, however: millions of hero-worshipers could not get enough of the daring explorer who had so ignited their imagination with his Gobi adventures. An undertaking of the magnitude of the Central Asiatic Expeditions epitomized certain prevalent attitudes of the 1920s: a taste for audacious ventures, a love of flamboyance and craving for excitement, a profound curiosity regarding the earth's last unknown geographical realms. In addition, there was a fanatical adoration of anyone identified with acts of personal courage—figures such as Charles Lindbergh, Richard Byrd, T. E. Lawrence, Lowell Thomas, Amelia Earhart, Richard Halliburton, and a host of other explorers, travelers, soldiers of fortune, and daredevils, who constantly made headlines with death-defying feats. With his dauntless, grandly conceived journeys into the Gobi and his defiance of conventional wisdom, Andrews embodied these popular preoccupations with incomparable panache, making his name a household word for millions.

Such was Andrews' celebrity that even the death in 1934 of his beloved German shepherd, Wolf, whom Andrews had been forced to leave behind in Peking with a friend, was reported in the *Washington Post* and other newspapers from a United Press wire service story:

> Peiping, China (U.P.)—After 13 years of adventuring in North China, Mongolia, and on the tablelands of Central Asia, Wolf, one of the world's most famous dogs, is dead. The constant companion of Dr. Roy Chapman Andrews on all the travels of the Central Asiatic Expeditions of the American Museum of Natural History, Wolf achieved a fame which extended far beyond the borders of China. . . .

Meanwhile, Andrews' responsibilities at the Museum were rapidly becoming intolerable. Although he tried to disguise it, he was painfully unhappy in his role as director. "It was inevitable," he once remarked, "that I should live a life that gave me the wild places of the world as a playground." One can readily understand, therefore, why he felt confined and ill at ease behind a desk, shackled with tedious administrative duties and a never-ending burden of monetary difficulties. Nature had predisposed Andrews, physically and psychologically, to seek out the earth's "wild places" with single-minded determination. Without access to the field, he had become what he described as "a square peg in a round hole"—a world-famous anomaly hopelessly cut off from the arena in which his talents had flourished so brilliantly.

Added to the emotional strain of his work, Andrews' health began to suffer from lack of exercise. "Never in my life had I been ill," he wrote, "but [now] every germ in New York City found me a delightful place for breeding." Desperately in need of physical activity and a place to escape the city on weekends, he and Billie set out to acquire a country retreat. Early in June of 1937, Billie lunched with a friend who specialized in rural properties. The following Saturday, Billie and Andrews drove out to see what was to become their country residence for the next sixteen years. Guided by a map drawn by the agent, they ended up in the Berkshires of northwestern Connecticut in the tiny village of Colebrook. Just past the town green, one mile up a steep hill, stood a sadly rundown house on 150 acres of pristine land, with a beautiful pond set jewel-like in a forest of pines, maple, silver birch, oak trees, and flowering shrubs. Even though everything needed renovation and tangles of underbrush choked the property, Billie bought the place the following Monday. Surveying the forest and the tranquil pond, she and Andrews named it Pondwood Farm. From then on, nearly every weekend and holiday was spent clearing the grounds, rebuilding the house, constructing duck blinds on a nearby lake, putting in gardens and a small rock-lined swimming pool. Later, they installed trails for skeet shooting, and erected a log cabin in the woods to provide Andrews with a studio for writing.

For the first four years, Andrews and Billie commuted the 125 miles to their apartment in New York on Sunday evenings and returned to Pondwood Farm on Friday afternoons. It was an ideal arrangement. But matters at the Museum gradually went from bad to worse. First came the shattering news that Walter Granger—Andrews' intimate friend and the stalwart second-in-command and scientific coordinator of the Central Asiatic Expeditions—had died of a heart attack on September 2, 1941, at age sixty-eight while on a fossil-collecting trip to Lusk, Wyoming. Andrews and a group of Granger's friends and associates were on hand at Pennsylvania Station to receive his body, and Granger's death sent shock waves far and wide. Andrews was inundated with letters of condolence from former staff members of the expeditions and colleagues in China.

Another blow fell after Pearl Harbor, when Trubee Davison took a leave of absence from the Museum to accept a position with the air force in Washington. "When Trubee left," wrote Andrews, "the one bright spot in my Museum work 'blacked out.' . . . I had become increasingly unhappy at the Museum. Raising money, trying to make both ends meet in our budget was the be-all and end-all of my existence. . . . My work might as well have been carried on from an office in Wall Street."

At this point, Billie, concerned with the effects of these pressures on her husband, was urging Andrews to resign, but he found it difficult to leave the institution that had occupied thirty-five years of his life. Yet he could no longer deny that neither by temperament nor training was he the right person to carry the Museum through its worsening crisis. "When it became evident that meeting the next budget would mean wholesale retirements of men like Frank Chapman, Clark Wissler, Barnum Brown and others who had made the institution great, but were along in years, I couldn't face it. So I offered my resignation to the Board to take effect January 1, 1942, and was appointed Honorary Director."

Andrews' motives for leaving, however, were not altogether altruistic. As discontent among the staff had continued to mount, the trustees invited Alexander G. Ruthven, the president of the University of Michigan and formerly a herpetologist, with considerable knowledge of museum management, to appraise the Museum's problems and make recommendations for their improvement. Aside from criticizing the antiquated design of the buildings, the inefficient use of space, and the disorganized display of collections, Ruthven particularly denounced the installation, during a time of financial stress, of large, expensive exhibits such as the Akeley Memorial Hall of African Mammals and the Hall of North American Mammals, two of the showpieces of Andrews' tenure. Ruthven went on to single out "a weak administrative policy [that] has resulted in faulty organization." His advice was unequivocal: strengthen the administration by "changes in personnel." Thus, even if Andrews felt the time was right for his departure, the deciding factor behind his resignation was undeniably the Ruthven Report, which the trustees unanimously adopted. "A younger man able to adjust himself and his ideas to the aftermath of the war," Andrews conceded, "was needed to direct the destiny of what I believe is one of the most worth-while institutions in all the world."

When Andrews told Billie of his decision to resign, she was deliriously happy. At last, he was free of the debilitating stress, constant social obligations, overly rich fund-raising banquets, and stinging criticism to which he had been subjected for seven years. It so happened that his son George was in New York on leave from the military that day, and they drank a toast to a new episode in the life of the aging, disillusioned explorer whose soul had never left the desert.

Whatever regrets Andrews harbored over leaving the Museum were soon dispelled by his new role as a "country squire." Life at Pondwood Farm was nothing less than euphoric. Once the renovation of the house was completed, Billie set about remodeling the interior. A few select Chinese antiques were mixed with comfortable furniture, and the walls of the dining room were painted Chinese red and embellished with Andrews' prized Ming Dynasty screens. A large picture window was installed in the living room, where Billie and Andrews frequently ate breakfast or lunch served on trays next to a blazing fire while watching the antics of the wildlife around the pond. Andrews converted the original living room into a pine-paneled gun room lined with books, framed medals and awards, and a glass-fronted cabinet filled with rifles, shotguns, fishing rods, and his cherished 6.5 Mannlicher, which he had carried on his expeditions. Every winter, fragrant logs roared in the stone fireplace, and whatever the season, it was the inviting gun room that invariably lured guests for cocktails. "It is primarily a livable house," wrote Andrews, "filled with light, color, and freedom where people, dogs and cats may roam at will." Billie and Andrews shared a passion for animals that included a menagerie of well-trained hunting dogs, several cats, a flock of ducks, pigs, and chickens.

A steady procession of friends was invited to Pondwood Farm for skeet shooting, hunting, or fishing: ducks, grouse, woodcocks, trout, and bass were especially plentiful. Afterward, everyone gathered at the house for drinks, lunches, or buffet suppers of freshly prepared game and fish, or gourmet summer barbecues in the rock garden. And occasionally George Andrews would arrive for the weekend from Princeton with friends or classmates.

Almost every day at Pondwood, Andrews retreated for several hours to his cabin in the woods to write. He worked at a large desk surrounded by bookcases and framed photographs of his fellow explorers: Peary, Byrd, Stefansson, Wilkins, Bartlett, Ellsworth, Granger, Younghusband, Amelia Earhart, Lindbergh, and others, autographed with personal messages.

Andrews was in constant demand as an author, and he turned out one article after another dealing with natural history, hunting, his Gobi expeditions, and exploration in general. He contributed features to *Natural History*, the *Literary Digest*, *Scientific American*, *Reader's Digest*, *American Magazine*, *Collier's*, the *San Francisco Chronicle Magazine*, *Cosmopolitan*, *Science Illustrated*, *Outdoor Life*, the *New York Times Magazine*, *Field and Stream*, *True: The Man's Magazine*, and other periodicals. In addition, he produced several more popular books, including his autobiography *Under a Lucky Star*, which he wrote at Billie's insistence after breaking his leg during the winter of 1942. For six weeks, he sat with his leg in a cast, wrapped in a fur-lined Chinese silk robe, writing in longhand or dictating to Billie, who then typed the manuscript.

The book appeared in 1943 and was an immediate best-seller. Its success inspired him to publish a sequel in 1947, *An Explorer Comes Home*, a charming chronicle of life at Pondwood Farm. Andrews also tapped into the lucrative market for books for young readers, some of which were enormously popular, especially his *Quest in the Desert*, *All About Dinosaurs*, *Quest for the Snow Leopard*, and *All About Whales*, each of which was inspired by his own travels.

Once a week, in the winter of 1942–1943, Andrews and Billie went by train or car to Manhattan, staying for two or three days at the Weylin Hotel. Despite hobbling about on crutches for his broken leg, Andrews completed his fifth and final year of broadcasts for Columbia's *School of the Air*, took care of business, and attended the Tuesday lunches of the Dutch Treat Club. Billie delighted in donning her elegant clothes—for which she was famous—to visit friends, shop, and accompany her husband to dinner, the theater, or the opera.

When the Japanese attacked Pearl Harbor, Andrews volunteered his services to the military, and although he was rejected as too old for active duty, he served as a consultant on Asian affairs and advised General George S. Patton on training troops for desert warfare. Andrews' son George, recently graduated from Princeton, was accepted as a cadet in the Army Air Corps and sent to Texas for training, where he met his future wife, Mary Nancy McElhannon, a native of Sherman, near Dallas. Because George spoke fluent French, he was kept in the United States for eighteen months after earning his commission as a second lieutenant, teaching Free French airmen to fly Thunderbolt fighters. Next, he was sent to Europe and flew forty-three combat missions in support of the Ninth Army. After the war, he and Mary Nancy eventually settled in Dallas, where he established a financial consulting firm.

George's brother, Kevin, meanwhile, fought as an infantryman in the Italian campaign, and after the war graduated summa cum laude from Harvard. Winning a postgraduate grant to the American School of Classical Studies in Athens, he later abandoned archaeology and traveled widely in Greece. He wrote a scholarly study entitled *Castles of the Morea* (1953), and a lively book called *The Flight of Ikaros* (1959), recounting his adventures among peasants, fishermen, and the guerrilla fighters opposing the notorious junta then in control of Greece. Kevin married Nancy Cummings, the daughter of the celebrated poet E. E. Cummings, and settled in a house in Athens overlooking the Acropolis. When they later separated, Kevin "went native," living as a Greek citizen in a peasant section of town where he earned a living by making jewelry and writing. He drowned in 1989, the apparent victim of an epileptic seizure, while swimming with an American friend between the islands of Avgó and Kíthira.

Years earlier, the question of Kevin's rightful father had taken a curious turn. As late as 1937, the year Billie acquired Pondwood Farm, Andrews was still clearly unaware that Kevin might not have been his son. This is substantiated by the fact that Kevin and Andrews saw each other as often as possible. Kevin visited Pondwood Farm on numerous occasions, and Andrews had paid for his education. One day, though, Yvette inexplicably appeared at Pondwood, engaged in a violent confrontation with Andrews behind closed doors, and took Kevin away with her. He would never again return. After he moved to Greece, Yvette then traveled to Athens and revealed to Kevin that his actual father was her English lover and Andrews' close friend in Peking, Chips Smallwood, whose photograph was found among Kevin's belongings after his death. His brother (or half-brother), George, remains somewhat unconvinced that Yvette's confession was true, citing a physical resemblance between Kevin and Andrews. Yet Richard S. Kennedy, the author of a biography of E. E. Cummings, *Dreams in the Mirror*, refers to Kevin unequivocably as "the stepson of Roy Chapman Andrews," a statement presumably based on interviews with Kevin and his estranged wife, Nancy Cummings. Whatever the truth, Yvette's actions came as such a blow to Kevin, who idolized Andrews, that he rejected his mother from that time onward, even refusing to help with her funeral arrangements. And Andrews' fury over the matter prompted him irrationally to shut the door on any future contact with Kevin. Never again did Andrews mention him in print or correspond with him, and his already strained relationship with Yvette became extremely bitter.

Having lived somewhat like a Gypsy after her divorce from Andrews, traveling incessantly and often staying with friends for long periods, Yvette—ever fascinating and enigmatic—worked for military intelligence during the war, using her fluent command of German to help break enemy codes. She died in 1959 in a single-car accident near Burgos, Spain, while touring with her cousin Patricia Emmet.

Life at Pondwood Farm had proven to be the ideal panacea for Andrews' restlessness, the scars left by his none-too-successful tenure at the Museum's helm, and, most painful of all, the devastating effects of the premature end of his adventures as an explorer after so many spectacular achievements. At Pondwood, he had found new friends to satisfy his gregarious nature, a closeness to the woodlands, streams, and lakes he had loved since his boyhood in Beloit, and time to reflect and find a thread of continuity in the violently contrasting mosaic of his career. And there was Billie, of course, whose beauty, self-confident resolve, and fun-loving warmth made her the perfect companion for this phase of Andrews' life.

"Oh, I am so happy!" Billie had exclaimed to Andrews one day at Pondwood. "And so we were and so we are," he added, ". . . to live again in the sunshine and the open air; to write when I want to write, to work in the good earth, to fish and shoot and swim, is Paradise enough. I feel as I did the night the *Albatross* rode into Keelung harbor, after battling a typhoon in the Formosa channel. My personal ship during thirty-five years has sailed all the oceans of the world; now at the end of its restless voyaging it has come to anchor in the quiet waters of Pondwood Pond."

But his "ship" would not remain there permanently. As Andrews grew older, the severe Berkshire winters drove them to seek a warmer winter climate. Having visited Tucson, Arizona, during their travels, they began spending every winter there from 1947 to 1954, basking in the sun, discovering Mexican food and margaritas, and studying the desert's wildlife and exotic vegetation.

Aside from the rigorous winters at Pondwood Farm, the physical demands of its upkeep were taking a toll on Andrews' stamina. At sixty-nine, he was no longer able to cope with trimming trees, cutting back underbrush, clearing trails, and various other tasks. After an agonizing evaluation of the situation, he and Billie reached a decision that would have been unthinkable a few years earlier. In 1954, they sold Pondwood to a resident of Stamford, Connecticut, Richard Haskell, and departed for the warmth of Tucson. Fascinated by the desert's flora and wildlife, Andrews served briefly on the board of the newly founded Arizona–Sonora Desert Museum. (After his death, Billie donated an endowment fund to the museum in Andrews' name to be used for research.) But the desert—beautiful as it was—proved unbearably hot except for the midwinter months, and searching for a more temperate climate, they moved to Carmel Valley in northern California. Here they purchased a charming white stucco and stone house with a superb view of the valley and surrounding hills. They instantly felt at home. Andrews wrote occasional articles, reviewed books, and spent much of his time shooting and conducting field trials for hunting dogs. Billie took up painting and gardening, and the combination of Andrews' fame and Billie's hospitality soon resulted in a busy social life.

Then the inevitable happened.

Andrews, now seventy-six years old, had fought a bout with lung cancer from which he was recovering. Shortly afterward, however, he began suffering chest pains and was hospitalized for observation and treatment. Four weeks later—on March 11, 1960—he died from a massive heart attack. Following his cremation, a memorial service was held in Carmel attended by a few close friends, and his ashes were returned to his family's plot at Oakwood Cemetery in Beloit.

Still only fifty-three years of age, Billie, never one to cling to grief, struggled to regain her equilibrium after twenty-five years of marriage to Andrews. In 1961, she remarried, this time to a successful radio executive, Robert A. Street, who owned stations on the West Coast. He had been a member of Andrews' and Billie's social circle, and was an accomplished seaman who cruised California's coast in an authentic Chinese junk made for him in Hong Kong. Their marriage lasted almost thirty-six years, and after living in San Diego, they moved back to Carmel Valley, settling in a hillside house among towering pines.

Soon after Billie suffered a paralyzing stroke early in 1997, Bob succumbed to stomach cancer. A second stroke left Billie bedridden, though her memory and lightning-quick wit remained intact almost until she passed away on September 21, 1998, at ninety-one. She died at home, surrounded by Andrews' collection of Chinese antiques—eerie reminders of his extraordinary life in the Orient.

Although widely reported in the world's press, Andrews' death was allowed to pass unnoticed in all of the American Museum's publications, leaving Billie, George, and droves of his friends understandably enraged. Only in recent years has a new generation of Museum officials atoned for such a glaring oversight by acknowledging Andrews' remarkable contributions to that institution's collections and reputation. No one today questions the fact that Andrews embarked on one of this century's most compelling and scientifically important ventures, an act of daring and imagination that assumed the quality of an epic drama played out in an exotic and dangerous setting. As Stephen Jay Gould observed in his book *Ever Since Darwin*: "The sheer romance [of the Central Asiatic Expeditions] fit Hollywood's most heroic mold." More important, Andrews revealed the Gobi Desert—so long dismissed by scientists as a wasteland—as one of the earth's richest repositories of extinct mammals and reptiles, a treasure trove whose fossil record has implications that extend far beyond its boundaries.

Even a cursory glimpse into his life reveals an unlikely mixture of entrepreneur, scientist, adventurer, and socialite. Rarely contemplative or given to self-doubt, Andrews was described by the writer D. R. Barton as "a fiery comet . . . [with] a sheer genius for finding a milieu wherein his problems can be solved by his great forte—direct, aggressive action." Adored by the public and ceaselessly pursued by the press, he came as close to superstar status as any explorer of this century. It has even been suggested by numerous writers that he served as the model for Hollywood's indomitable archaeologist-adventurer Indiana Jones. Although this assertion has been repeatedly denied by the character's creator, George Lucas, the analogy is not completely far-fetched. Andrews has also been compared to explorers of the Victorian era, and it is true that his outlook owed much to those illustrious nineteenth-century adventurers—Baker, Stanley, Burton, Doughty, and the like—who undertook prodigious quests with unwavering resolve, at great personal risk, and without the benefit of today's high-tech support systems. Like most explorers, Andrews was stimulated by danger. He relished the opportunity to test himself against hardships, and he was willing to push his luck to the outer limits to gain his objectives.

Always the consummate adventurer, Andrews was driven by an addiction to distant lands and the lure of discovery. He was a "field man" par excellence: intrepid, resourceful, courageous, a brilliant organizer, and a skilled leader. He once wrote, "Exploration is my passion. Nothing equals the fulfillment it gives me. . . . I respond to its challenges with an anticipation unmatched by anything else." Such a statement was not meant lightly. The record of Andrews' attainments reflects its impact upon his priorities. His supreme accomplishment—the Central Asiatic Expeditions—represents its ultimate manifestation. "If I've left anything worthwhile to posterity," he once commented in response to a reporter's question, "it is the legacy of my Gobi journeys. I hope that is the yardstick by which my success or failure in life will be judged."

Rightly so, history has granted that wish. Long before his death in 1960, his name had become inseparably linked with the Gobi Desert, its scientific treasures, and the epic adventures of modern times. Together with men such as Peary, Scott, Amundsen, Hedin, MacMillan, and Byrd, Andrews ranks in the forefront of twentieth-century explorers, joining those select few who conquered the earth's last unknown regions.

EPILOGUE

In retrospect, Andrews' explorations in the Gobi ended on an ironic note. He had undertaken his quest hoping to prove Osborn's theory that central Asia had witnessed the genesis and dispersal of mammals, with emphasis on the search for mankind's precursor—the famous "Missing Link." His efforts, however, failed to turn up any traces of human fossils: not a splinter of bone or a single tooth remotely comparable in age to the relatively advanced *Homo erectus*, represented in Asia by Peking and Java Man. Although Andrews left China in 1932 still echoing Osborn's conviction that Asia would ultimately prove to be the "Cradle of Mankind," years later, he was forced to accept the fact that Africa would earn that distinction after all. Yet his failure to recover early anthropoids was overshadowed by the history-making revelation that the Gobi was one of the richest repositories of paleontological treasures on earth.

Andrews and Granger often experienced the frustration of leaving behind skeletons of beasts too large to be excavated in the short time available to them each season. As it was, the tons of specimens collected and the discovery of lucrative deposits for future study were astonishing feats given the region's awesome logistical and political problems. Nor was Andrews able to penetrate the enormous expanses of the western and southern Gobi, which he correctly suspected held vast graveyards of extinct animals. Moreover, many of the sites discovered by his expeditions have since proven inexhaustible. The desert conceals such a wealth of fossils that the relentless process of erosion exposes a bountiful harvest of new specimens every few years.

Not only is the Gobi littered with a mind-boggling array of Cretaceous and Cenozoic animals, but most of them are found in a remarkable state of preservation. Everything from tiny skulls of ancient mammals to skeletons of dinosaurs that roamed this area 90 million to 65 million years ago frequently emerge from the earth looking as if they had been buried only a few decades ago. Various theories have been advanced to account for this phenomenon. One of the most plausible was suggested by four paleontologists from the American Museum—Michael J. Novacek, Mark Norell, Malcolm C. McKenna, and James Clark, all of whom have worked extensively in the Gobi. Writing in the December 1994 issue of *Scientific American,* they speculated that the Gobi differed from other fossiliferous areas in various parts of the world, which were semitropical and cut by numerous streams and swiftly flowing rivers that scattered animal remains over large areas. By contrast, the late Cretaceous environment of the Gobi was much like that of the present day—a relatively dry maze of sand dunes, eroded badlands, wide valleys, and cliffs sparsely watered by seasonal lakes and streams. As there is considerable evidence in the Gobi's geological strata of violent sandstorms similar to those that sweep across the region today, such storms, as well as mud slides caused by brief torrential rains, could very likely have buried the carcasses of animals or suffocated living species under thick sand, thereby protecting their bodies from scavengers and leaving them practically undisturbed.

No sooner had World War II ended before the widespread interest ignited by Andrews' discoveries touched off renewed paleontological explorations in the Gobi. By 1946, with Outer Mongolia firmly in Russia's grip, a series of three expeditions was launched under the auspices of the Academy of Sciences in the Soviet Union, Organized by I. Orlov and led in the field by I. Efremov and A. Rozhdestvensky, all highly respected paleontologists, the venture was reminiscent of Andrews' undertaking. Although the supply caravans and antiquated Dodge automobiles were replaced with heavy-duty military trucks and five "scouting cars" similar to jeeps, the Russian party consisted of a sizable group of scientists, technicians, mechanics, laborers, and cooks, all of whom were capable of investigating several localities at once.

During 1946, 1948, and 1949, the Russians prospected at a place called Bain Shirch, southeast of Ulan Bator, then traveled westward to the Flaming Cliffs, where they recovered more *Protoceratops* skeletons, eggs, and a grotesque armored creature called *Syromsaurus.* With their sturdy trucks, they were able to navigate a route south and west of the Altai Mountains into the forbidding terra incognita that Andrews had only glimpsed from a distance. Here, the Russian explorers forged their way slowly through a hellish, sweltering depression known as the Nemegt Valley, a jumble of multicolored badlands and amphitheaters flanked on the north and south by jagged mountains as high as ten thousand feet. But they were spectacularly rewarded for their efforts: the Nemegt Valley was a fossil hunter's paradise. Everywhere lay massive bones of duck-billed dinosaurs called hadrosaurs, lumbering armor-plated ankylosaurs, and skeletons of *Tarbosaurus*, a huge carnivorous predator virtually identical to *Tyrannosaurus rex*, which once roamed the western United States. Before the expedition ended, the Russian teams had brought 120 tons of fossils out of the Gobi, established the Nemegt Valley as the focal point for future explorations, and verified Andrews' predictions regarding the paleontological treasures still hidden beneath the desert.

Next came an extended series of joint Polish-Mongolian expeditions led by a remarkable woman named Zofia Kielan-Jaworowska. Over the course of eight field seasons—from 1963 to 1971—she and her colleagues embarked on an ambitious program of exploration that carried them to sites from Shabarakh Usu through the Nemegt Valley into even more distant regions to the north. The inventory of fossils uncovered by Kielan-Jaworowska and her Polish and Mongolian colleagues was nothing less than incredible: superbly preserved tarbosaurs, sauropods, hadrosaurs, *Protoceratops*, ankylosaurs, and *Velociraptor*, the terrifying villain of *Jurassic Park* fame. One of the most astounding discoveries made by the Polish-Mongolian team was the so-called "fighting dinosaurs." Excavated at Tugrugeen Shireh, thirty miles west of the Flaming Cliffs, were the nearly perfect skeletons of a *Protoceratops* and a *Velociraptor* locked in mortal combat. The *Protoceratops* was clutching its enemy's right arm in its jaws, while the *Velociraptor* tore away at its prey's neck with lethal sicklelike "killer hooks," especially on the middle toe of the *Velociraptor*'s hind feet. Now one of the prized possessions of the Natural History Museum in Ulan Bator, the two beasts, embedded in sandstone, lie frozen in their death struggle just as they were unearthed. Added to Kielan-Jaworowska's other bounty was a valuable collection of lizard skulls and dozens of small mammals from

Velociraptor

The fierce killer of Jurassic Park *fame is notable for its slashing teeth and sicklelike rear claws. Like the oviraptorids, this theropod exhibited avian traits, especially thick feathers.*

Shabarakh Usu, Tugrugeen Shireh, and a section of the Nemegt Valley appropriately named "Eldorado" by the explorers because of its paleontological wealth.

In the years between 1986 and 1990, another of several more multinational expeditions—one of the largest ever assembled—took to the field in both Inner Mongolia and northwestern Canada, in an attempt to resolve a compelling theoretical question that harked back to a concept originally proposed by Andrews, Granger, Osborn, and Matthew. It involved possible relationships between Asian dinosaurs and related types found in the western United States and Canada. In an article in *Earth Science* in 1989, Philip J. Currie of the Tyrrell Museum of Paleontology in Drumheller, Alberta, summarized this hypothesis as first hinted at by the Central Asiatic Expeditions' discoveries: "Among their many significant finds," Currie observed, "was evidence to suggest land connections that allowed animals to move between North America and Asia as early as the Cretaceous period. Although the Cretaceous faunas of the northern continents are not identical, most families of dinosaurs from that time have representatives in both Asia and North America.

With this fact in mind, Canadian scientists joined forces with their Chinese counterparts to launch the Sino-Canadian Dinosaur Project, a five-year venture organized by the Institute of Paleontology and Paleoanthropology in Beijing, the National Museum of Natural Sciences in Ottawa, and the Tyrrell Museum of Paleontology in Alberta. Funding was provided by the Chinese Academy of Sciences, the Canadian government, and the Ex Terra Foundation in Alberta. Included in the expeditions were paleontologists, geologists, climatologists, and other specialists, supported by technicians, field assistants, and excavators. Explorations were carried out simultaneously in Inner Mongolia and remote areas of Alberta and Canada's Arctic islands—regions once linked by ancient land bridges over which dinosaurs could have migrated.

With the expeditions divided into several teams working at widely scattered locations, the yield of fossils was exceptionally rewarding. It included at least a dozen genera of dinosaurs, such as the familiar hadrosaurs, *Velociraptor*, *Protoceratops*, tyrannosaurids, ankylosaurs, and *Oviraptor* (the "egg thief" found by Andrews' party at the Flaming Cliffs apparently raiding a nest of *Protoceratops* eggs), together with lizards, turtles, and a variety of small mammals. One especially interesting site, Bayan Manduhu, produced a cluster of five *Protoceratops* skeletons lying on the side of a sand dune with their noses pointed upward as if gasping for air as a sandstorm smothered them. An equally dramatic tragedy in the same region left the bodies of five juvenile *Pina-*

Ankylosaur

Four species of ankylosaur have been discovered in the Gobi. Often weighing several tons and reaching over twenty feet in length, they were covered by layers of protective armor, and their spiked tails provided a lethal weapon against predators.

cosaurus, a genus of the armored ankylosaurs, similarly trapped under a dune, the apparent victims of suddenly shifting sand. As suggested earlier, these multiple deaths, not uncommon in the region, offered persuasive evidence that large numbers of the Gobi's Cretaceous denizens perished and were quickly buried in swirling infernos of sand that left their remains amazingly preserved.

One of the historic moments of the Sino-Canadian explorations occurred when Philip Currie, a principal organizer of the project, attempted to locate Iren Dabasu, where Andrews' expedition had found the first dinosaur bones in central Asia in 1922. Uncertain that the spot he had pinpointed was actually Iren Dabasu—as landmarks and placenames had changed drastically since the Americans had worked there so long ago—Currie was elated when his Chinese colleague, Dong Zhiming, dug into a mound and produced a crushed metal flask engraved with the image of a family driving a 1920s Dodge touring car. Below it the word "Spirits" was plainly visible. The curious object was one of the hip flasks specially designed by Dodge Brothers and presented to every member of Andrews' expeditions as a token of the company's support.

After their five years of work in both hemispheres, the Canadians and Chinese were able to demonstrate that large sections of the continental masses were not covered by oceans during the late Cretaceous and were connected in their northernmost latitudes by land bridges. As Michael J. Novacek has pointed out, a fairly wide variety of dinosaurs made use of these intercontinental links to travel between Asia and North America. For example, the hadrosaur *Saurolophus* is found in both regions; *Velociraptor,* ankylosaurs, and *Oviraptor* had close relatives in North America, as did *Tarbosaurus,* a slightly smaller version of *Tyrannosaurus,* which are so much alike that some paleontologists believe they should both be classified as a single species. And a theropod found in the Gobi, *Saurornithoides,* has a counterpart known as *Troodon* that inhabited Alberta, Wyoming, and Montana.

With the end of the Sino-Canadian explorations in 1990, Currie felt confident in stating, "There can be no doubt that during the Late Cretaceous (between 65 million years and 80 million years ago), animals were moving freely between Asia and North America." Currie added that such intercontinental migrations may have gone in both directions, "since almost every family of dinosaurs," he wrote, "known from Cretaceous rocks in North America also had representatives in Asia." Even though it will require years of laboratory analysis and additional fieldwork to fill the gaps in this scenario, the Sino-Canadian project yielded overwhelming evidence that at some point during the late Cretaceous, Asian and North American dinosaurs had strong evolutionary affinities.

One of the outgrowths of the postwar expeditions that brought scientists back to the Gobi in search of further insights into its fossil record was the enactment of protocols that facilitated multinational cooperative efforts. With the emergence of Mongolia and China as sovereign nations, laws were passed that clearly established each country's ownership of their paleontological resources. As a result, the kind of ambiguous, politically charged situations Andrews had encountered with the Cultural Society in China, and before that during the Communist takeover of Outer Mongolia, were eliminated. In addition, Chinese and Mongolian universities, museums, and scientific societies began turning out well-trained paleontologists, geologists, and other specialists capable of undertaking explorations on their own, or joining forces with colleagues from other countries to initiate research projects on an equal footing. These innovations ushered in a new era of international cooperation that paved the way for American scientists to retrace Andrews' footsteps in the Gobi seventy-five years after the Central Asiatic Expeditions were forced to leave Outer Mongolia in 1925.

Although the American Museum of Natural History had been trying for years to gain entry into the Gobi, their request was never granted until the Cold War barriers came down in 1990. At the invitation of the Academy of Sciences in the Mongolian People's Republic, three paleontologists from the Museum—Michael J. Novacek (the Museum's vice-president and provost of science), Mark Norell, and Malcolm C. McKenna (both curators of vertebrate paleontology)—were invited to visit the country. Accompanied by two Mongolian scientists and traveling in Russian army trucks, they made a two-week reconnaissance of the Flaming Cliffs and the Nemegt Valley.

Along with a promising collection of specimens, a sensational discovery awaited them at a site in the Nemegt Valley known as Khulsan. Here, Malcolm McKenna retrieved an eight-inch skull and assorted bones of a previously unknown lizard. Originally seven feet in length, with knifelike teeth, it was, to quote Mark Norell, "a dead ringer" for the famous Komodo "dragons," flesh-eating lizards found today in Indonesia. Except for peculiar canals in its teeth, possibly designed for injecting venom into its prey, it appeared to be closely related to the same family as the Komodo—the Varanidae. Slight differences in the eye sockets, however, showed that it was a different genus; accordingly, the scientists named it *Estesia* in honor of a colleague, Richard Estes, the world's leading expert on fossil lizards who had died a year earlier.

Based on the results of this preliminary reconnaissance, an agreement was drawn up in Ulan Bator between the American Museum and Mongolia's Academy of Sciences for cooperative explorations that began in 1991. Under this protocol, Novacek, Norell, and McKenna were designated as the principal representatives for the American Museum. The Mongolian government assigned three of its leading paleontologists to accompany the expeditions—Demberelyin Dashzeveg, Altangerel Perle, and Rinchen Barsbold.

In the summer of 1991, the Americans returned to the Gobi equipped with three four-wheel-drive Mitsubishi Monteros and an unreliable Russian truck for transporting fossils. Added to camping equipment, collecting paraphernalia, and food, they brought along high-tech items such as laptop computers, a navigating device called GPS (global positioning system) that could fix the explorers' location by signals received from earth-orbiting satellites, and remote sensing images transmitted by LANDSAT and SPOT satellites. Novacek served as leader for the overall planning, funding, and administration of the project; Norell directed field operations, and McKenna was in charge of logistics, navigation, and communications. McKenna's wife, Priscilla, was responsible for making a continuous road log based on GPS readings. Over the years, specialists in various branches of paleontology, geology, and other disciplines were added to the Mongolian Academy–American Museum Expedition (MAE) to study new fossils, strata, and scientific problems. To review all the discoveries resulting from these renewed explorations is beyond the scope of this epilogue. They have been superbly documented in Novacek's book *Dinosaurs of the Flaming Cliffs* (Anchor Books, 1996), as well as in numerous articles in *Nature, Science, Scientific American, National Geographic,* and *Natural History.*

Eventually, the American Museum–Mongolian expeditions reaped an extraordinary harvest of fossils. During the first two seasons, however, the explorers' progress—especially in 1991—was severely hampered by unanticipated problems resulting from the collapse of Mongolia's infrastructure following its independence from Russia. Gasoline was in short supply, and food shortages became a serious obstacle. They were unable to purchase many staples available in Ulan Bator only a year before, forcing the scientists to survive on little more than freeze-dried rations brought from the United States to supplement their diet. In 1992, the difficulty of replenishing their supply of gasoline was solved by the acquisition of a cumbersome tanker truck loaded with thirteen hundred gallons of 76-octane gas for the Russian vehicles, and a second tanker, attached to the larger one, carrying 780 gallons of 93-octane fuel for the Mitsubishis. The food crisis was likewise resolved by large shipments of staples, delicacies, and wine sent from Los Angeles.

At last, in 1993, the explorers hit the bonanza they had been seeking for two years. Entering the Nemegt Valley, they traveled northward into an area of undramatic reddish-brown hills situated in an amphitheater-like basin known locally as Ukhaa Tolgod (Brown Hills). Within hours after arriving, they were dumbfounded by the extravagance of fossils gleaming in the scorching sun. "Like Granger and Andrews when they discovered [the] Flaming Cliffs," wrote Novacek, "we couldn't believe that our encounter with this fossil-laden patch of ground was real. Things like this don't happen, even to paleontologists on a major expedition like ours. . . . We had achieved one of the primary goals of the expedition—finding a new locality, one that measured up to some of the well-known dinosaur sites of the Gobi, one with an abundance of fossils. But this was beyond anything we could have imagined. In an area the size of a football field we had found a treasure trove that matched the cumulative riches of all the other famous Gobi localities combined."

Scores of dinosaur skeletons emerged from Ukhaa Tolgod's brown earth, including many unclassified varieties in nearly pristine condition. Nests of eggs lay everywhere, intermingled with lizards, turtles, and a gold mine of assorted mammals. One block of sandstone found in 1994 contained six shrewlike placental mammals with the skulls still connected to the fragile skeletons. No one could believe the wealth of fossils that kept turning up at Ukhaa Tolgod, or what new insights they would reveal once the painstaking job of laboratory analysis was completed.

So far, the most widely publicized discovery at Ukhaa Tolgod has revolutionized our knowledge of the dinosaur eggs first unearthed at the Flaming Cliffs. It occurred when Mark Norell happened upon a specimen that has since altered the long-accepted belief that the eggs were laid by *Protoceratops*, the ubiquitous beasts that swarmed over the Gobi's Cretaceous landscape. Norell's disclosure immediately brought to mind the day in 1923 when George Olsen found a nest of what were presumed to be *Protoceratops* eggs on top of which lay a four-inch layer of sand and the skeleton of a previously unidentified dinosaur. Assuming that the intruder was a predator killed by a sandstorm in the act of raiding the nest, Osborn had named it *Oviraptor philoceratops* or "egg thief that loves ceratopsians," though he added with a note of caution that this designation could "entirely mislead us as to its feeding habits and belie its character." Osborn's caveat was prophetic, as Norell's discovery would demonstrate.

What Norell found at Ukhaa Tolgod was a cluster of eggs identical to those supposedly laid by *Protoceratops*. But the top of one egg had been worn away by erosion, exposing a well-developed embryo encased in the shell. To everyone's astonishment, the embryo was *not* that of a *Protoceratops*. Instead, closer examination showed the embryo's skeleton belonged to an *Oviraptor*, leading to the inescapable conclusion that the eggs so long attributed to *Protoceratops* were actually laid by *Oviraptor*. Osborn had been right in warning that the name "egg thief" might be misleading regarding *Oviraptor*'s feeding habits. All indications now show that the specimen from the Flaming Cliffs had not died while raiding the eggs found by Olsen, but rather while incubating them. As James Clark wrote in *Natural History*, "The 'egg thief' from the Flaming Cliffs was more likely to have been sitting on the eggs than preying on them."

Recent discoveries at Ukhaa Tolgod and Turgrugeen Shireh (where the famous "fighting dinosaurs" were found by the Polish expedition) have also shed light on the evolution of a group of dinosaurs, the theropods, and their living descendants, which we know as birds. Skeletons of birdlike theropods found by Malcom McKenna at Turgrugeen Shireh and later at Ukhaa Tolgod—bizarre creatures later named *Mononykus*—appear more like flightless birds than dinosaurs, although they retained definite reptilian characteristics. About the size of a turkey, it had long thin legs and claws, a skull with a birdlike beak, and stubby forearms equipped with a single claw. Its skeleton exhibited a pronounced "keel" on the breastbone, a trait strikingly similar to the skeletal structure of birds. Recent discoveries have likewise verified that two other theropods—*Oviraptor* and *Velociraptor*—exhibited unmistakable birdlike traits, including rudimentary feathers covering most of their bodies.

Among the fossils from Ukhaa Tolgod, as well as other regions examined by the Americans and Mongolians, were quantities of small mammals similar to those picked up by Granger at the Flaming Cliffs and since found by the hundreds throughout the Gobi. Altogether they comprise a priceless reference collection of Cretaceous marsupials and placentals—the direct ancestors of the diverse mammals that evolved during the Cenozoic era, the Age of Mammals. Moreover, the existence of large populations of Cretaceous mammals in the Gobi bears directly on Osborn's theory of mammalian origins and dispersal, the concept that inspired Andrews to launch the Central Asiatic Expeditions. But exactly how these miniature creatures, scurrying around under the feet of dinosaurs, evolved into the animals that influenced Osborn's thinking is shrouded in aeons of darkness.

Nor was the Gobi the earth's only theater of mammalian evolution. Mammals of comparable age have been found in various parts of the globe, including North America. A staggering amount of research remains to be done if we are ever to fill the void between the existence of Cretaceous mammals and the emergence of the huge assortment of mammals—extinct and living—that trod the earth following what paleontologists refer to as "the great extinction event," an environmental catastrophe that wiped out the dinosaurs. Given the tremendous gaps in our understanding of how and when multi-million-year-old marsupials and placentals no larger than shrews transformed themselves into the huge array of mammals that came to dominate the earth, Osborn's speculations must remain in limbo.

While Andrews failed in his quest for ancient man and provided no answers to his mentor's hypothesis, his explorations opened the curtain on a gigantic stage of ancient life frozen in the Gobi's rocks and sandy wastes. The Central Asiatic Expeditions—hampered by war, politics, and bandits, and working in total isolation without the benefit of modern technology—will always remain one of the grand adventures in the annals of exploration. No scientist who has since challenged the Gobi's awesome expanses, or who will ever do so in the future, can fail to recall the ghostly images of Andrews' thin line of supply-laden camels and spindly wheeled cars moving like ants across that engulfing place at the ends of the earth. In terms of romance, daring, and sheer audacity, we will never see the equal of his grand adventure again.

Selected Bibliography

Books by Roy Chapman Andrews
Whale Hunting with Gun and Camera. New York: D. Appleton and Company, 1916.
Camps and Trails in China. (With six chapters by Yvette Borup Andrews.) New York: D. Appleton and Company, 1919.
Across Mongolian Plains. New York: D. Appleton and Company, 1921.
On the Trail of Ancient Man. New York: G. P. Putnam's Sons, 1926.
Ends of the Earth. New York: G. P. Putnam's Sons, 1929.
The New Conquest of Central Asia. New York: American Museum of Natural History, 1932.
This Business of Exploring. New York: G. P. Putnam's Sons, 1935.
This Amazing Planet. New York: G. P. Putnam's Sons, 1937.
Under a Lucky Star. New York: The Viking Press, 1943.
An Explorer Comes Home. New York: Doubleday Company, 1947.
Heart of Asia: True Tales of the Far East. New York: Duell, Sloan and Pearce, 1951.

Articles by Roy Chapman Andrews
"Notes Upon the External and Internal Anatomy of *Balaena glacialis* Bonn." *Bulletin of the American Museum of Natural History,* vol. 24, 1908.
"Whale Hunting as It Is Now Done." *World's Work,* December 1908.
"A Summer with the Pacific Coast Whalers." *American Museum Journal,* vol. 9, no. 2, 1909.
"Around the World for the Museum." *American Museum Journal,* vol. 11, no. 1, 1911.
"Shore-Whaling: A World Industry." *National Geographic Magazine,* vol. 22, no. 5, 1911.
"An Expedition in Korea." *American Museum Journal,* vol. 12, no. 6, 1912.

"An Exploration of Northeastern Korea." *Natural History*, vol. 12, no. 7, 1912.

"The Wilderness of Northern Korea." *Harper's Magazine*, vol. 126, no. 756, 1913.

"The California Gray Whale (*Rhachianectes glaucus* Cope). Its History, Habits, External Anatomy, Osteology and Relationship." *Memoirs of the American Museum of Natural History* (new series), vol. 1, part 5, 1914.

"The Sei Whale (*Balaenoptera Borrealis* Lesson). Its History, Habits, External Anatomy, Osteology and Relationships." *Memoirs of the American Museum of Natural History* (new series), vol. 1, part 6, 1914.

"Little Known Mammals from China." *American Museum Journal*, vol. 17, no. 8, 1917.

"China's Ancient Monuments." *American Museum Journal*, vol. 18, no. 4, 1918.

"Traveling Toward Tibet." *Harper's Magazine*, vol. 136, no. 815, 1918.

"The Blue Tiger." *American Museum Journal*, vol. 18, no. 5, 1918.

"The Frontier of the Forbidden Land." *Harper's Magazine*, vol. 136, no. 816, 1918.

"Camps in China's Tropics." *Harper's Magazine*, vol. 137, no. 817, 1918.

"Zoological Explorations in Yunnan Province, China." *Geological Review*, vol. 6, no. 1, 1918.

"Exploring Unknown Corners of the 'Hermit Kingdom.'" *National Geographic Magazine*, vol. 36, no. 1, 1919.

"Across Mongolia by Motor-Car." *Harper's Magazine*, vol. 139, no. 829, 1919.

"A New Search for the Oldest Man: A Great American Expedition to Asia." *Asia*, vol. 20, no. 10, 1920.

"Big Game Hunting at the Eastern Tombs." *Asia*. vol. 20, no. 8, 1920.

"Urga: The Sacred City of the Living Buddha." *Harper's Magazine*, vol. 141, no. 842, 1920.

"The Lure of the Mongolian Plains." *Harper's Magazine*, vol. 141, no. 844, 1920.

"New Expedition to Central Asia: To the Earth's Most Ancient Center of Human Dispersal." *Natural History*, vol. 20, no. 4, 1920.

"In Mongolia and North China." *Natural History*, vol. 20, no. 4, 1920.

"Digging for the Roots of Our Family Tree." *Asia*, vol. 21, no. 5, 1920.

"Hunting the Great Ram of Mongolia." *Harper's Magazine*, vol. 142, no. 849, 1921.

"The Motor Truck in Central Asia." *Natural History*, vol. 21, no. 1, 1921.

"Politics and Paleontology." *Asia*, vol. 22, no. 5, 1922.

"The Quest for the Golden Fleece. I. The Wilds of Shensi." *Asia*, vol. 22, no. 6, 1922.

"The Quest for the Golden Fleece. II. Takin on Their Rugged Peaks." *Asia*, vol. 22, no. 7, 1922.

"Scientific Work in Unsettled China." *Natural History*, vol. 22, no. 3, 1922.

"The Third Asiatic Expedition of the American Museum of Natural History." *Science*, vol. 55, no. 1431, 1922.

"Gobi Bound." *Saturday Evening Post*, November 10, 1923.

"Untying Red Tape in Urga." *Asia*, vol. 23, no. 6, 1923.

"Tenting in Lama Land." *Asia*, vol. 23, no. 7, 1923.

"A Kentucky Derby in the Gobi Desert." *Asia*, vol. 23, no. 8, 1923.

"A Fossil-Hunter's Dream Come True." *Asia*, vol. 23, no. 10, 1923.

"Where the Dinosaur Hid Its Eggs." *Asia*, vol. 24, no. 1, 1924.

"The Lure of Mongolia." *Asia*, vol. 24, no. 2, 1924.

"Eggs at $60,000 a Dozen." *Saturday Evening Post*, May 24, 1924.

"'Mongols' of Twenty Thousand Years Ago." *Asia*, vol. 26, no. 1, 1926.

"The Colossus of Mongolia." *Asia*, vol. 26, no. 2, 1926.

"The Crowning Discovery in the Gobi." *Asia*, vol. 26, no. 4, 1926.

"On the Trail of Ancient Man." *World's Work*, February 1926.

"Explorations in Mongolia: A Review of the Central Asiatic Expeditions of the American Museum of Natural History." *Geographical Journal* (London), vol. 69, no. 1, 1926.

"Further Adventures of the American Men of the Dragon Bones." *Natural History*, vol. 29, no. 2, 1929.

"Gobi Death Traps." *Saturday Evening Post*, October 25, 1930.

"The Fate of the Rash Platybelodon." *Natural History*, vol. 31, no. 2, 1931.

"J. McKenzie Young." *Natural History*, vol. 32, no. 3, 1932.

"Explorations in the Gobi Desert." *National Geographic Magazine*, vol. 63, no. 6, 1933.

"The Mongolian Wild Ass." *Natural History*, vol. 33, no. 1, 1933.

Field Journals: 1919–1930. Rare Book Room, American Museum of Natural History.

Secondary Sources

Barton, D. R. "Gambler on the Gobi: The Story of Roy Chapman Andrews." *Natural History*, vol. 45, no. 2, 1940.

Behr, Edward. *The Last Emperor*. New York: Bantam Books, 1987.

Bodry-Sanders, Penelope. *Carl Akeley: Africa's Collector, Africa's Savior*. New York: Paragon House, 1991.

Botjer, George. *A Short History of Nationalist China, 1919–1949*. New York: G. P. Putnam's Sons, 1979.

Bredon, Juliet. *Peking: A Historical and Intimate Description of Its Chief Places of Interest*. Shanghai: Kelly & Walsh, 1922.

Caldwell, Harry R. *The Blue Tiger*. New York: Abingdon Press, 1924.

Clark, James L. *Good Hunting: Fifty Years of Collecting and Preparing Habitat Groups for the American Museum of Natural History*. Norman: University of Oklahoma Press, 1966.

Clark, James M. "An Egg Thief Exonerated." *Natural History*, vol. 104, no. 5, no. 6, 1995.

Clubb, Edmund O. *20th Century China*. 3rd ed. New York: Columbia University Press, 1978.

Colbert, Edwin H. *Men and Dinosaurs*. New York: E. P. Dutton & Co., 1968.

———. *Digging into the Past: An Autobiography*. New York: Dembner Books, 1989.

Currie, Philip J. "Long Distance Dinosaurs." *Natural History*, vol. 98, no. 6, 1989.

———. "Dragons and Dinosaurs." *Earth Science*, Summer, 1989.

———, ed. "Results from the Sino-Canadian Dinosaur Project." *Canadian Journal of Earth Sciences*, vol. 30, nos. 10–11, 1993.

Froelich, Louis D. "Andrews of Mongolia." *Asia*, vol. 24, no. 8, 1924.

Gallenkamp, Charles. "Roy Chapman Andrews and the Search for Early Man in Asia." *Exploration*. Santa Fe: School of American Research, 1978.

Goldstein, Melvyn C., and Cynthia M. Beall. *The Changing World of the Nomads*. Berkeley: University of California Press, 1994.

Hedin, Sven. *My Life as an Explorer*. Alfhild Huebsch, trans. New York: Garden City Publishing Co., 1925.

———. *Across the Gobi Desert*. H. J. Cant, trans. London: George Routledge and Sons, 1931.

Hellman, Geoffrey. *Bankers, Bones and Beatles: The First Century of the American Museum of Natural History*. Garden City: The Natural History Press, 1968.

Hopkirk, Peter, *Foreign Devils on the Silk Road*. Amherst: University of Massachusetts Press, 1980.

Kendall, Laurel. "The Devil Fish and the Tiger." *Natural History*, vol. 97, no. 5, 1988.

Kennedy, Richard S. *Dreams in the Mirror: A Biography of E. E. Cummings*. New York: Liveright Publishing Corporation, 1980.

Kielan-Jaworowska, Zofia. *Hunting for Dinosaurs*. Cambridge: The MIT Press, 1969.

Lattimore, Owen. *Mongol Journeys*. New York: Doubleday, Doran and Co., 1941.

Maclean, Fitzroy. *To the Back of Beyond.* Boston: Little, Brown & Company, 1975.

Morden, William J. *Across Asia's Snows and Deserts.* New York: G. P. Putnam's Sons, 1927.

Norell, Mark, A. *All You Need to Know About Dinosaurs.* New York: Sterling Publishing, 1991.

———, L. Chaippe, and J. Clark. "New Limb on the Avian Family Tree." *Natural History,* vol. 102, no. 9, 1993.

———, E. S. Gaffney and L. Dingus. *Discovering Dinosaurs.* New York: Alfred A. Knopf, 1995.

Novacek, M. J., M. A. Norell, M. C. McKenna, and J. Clark. "Fossils of the Flaming Cliffs." *Scientific American,* vol. 271, no. 6, 1994.

———. *Dinosaurs of the Flaming Cliffs.* New York: Doubleday, 1996.

Ossendowski, Ferdinand. *Beasts, Men and Gods.* New York: E. P. Dutton & Co., 1922.

Pond, Alonzo. *Andrews: Gobi Explorer.* New York: Grosset and Dunlap, 1972.

Pope, Clifford H. "A Snake Buyer in China." *Asia,* vol. 27, no. 8, 1927.

———. *China's Animal Frontier.* New York: The Viking Press, 1940.

Preston, Douglas J. *Dinosaurs in the Attic: An Excursion Into the American Museum of Natural History.* New York: St. Martin's Press, 1986.

Wallace, Joseph. *The American Museum of Natural History's Book of Dinosaurs and Other Ancient Creatures.* New York: Simon & Schuster, 1994.

Index